Essays on Inclusive Growth and Some Reminiscences

C.H. Hanumantha Rao is presently Chancellor of University of Hyderabad (Hyderabad Central University); Chancellor, ICFAI Foundation for Higher Education (Deemed to be University), Hyderabad; and Honorary Professor at the Centre for Economic and Social Studies (CESS), Hyderabad, as well as at the Institute of Economic Growth (IEG), University of Delhi.

Born on 15th May, 1929, in Karimnagar (Andhra Pradesh), he did his M.A. from the Osmania University, Hyderabad, in 1957, and Ph.D from the University of Delhi in 1962. He pursued his post-doctoral work at the Department of Economics, University of Chicago, during 1966-67.

He was employed as a Fellow at the IEG in 1961, from where he retired in 1992. He served as the Director of the Institute (1976-1980) and also as its Chairman (2002-2008). A Founder-Member of the CESS, he served as its Chairman during 1995-2007.

Prof Rao supervised, individually or jointly, a number of scholars for their Ph.D. theses submitted to the University of Delhi. He published 9 books, 7 edited books and over 100 research papers in academic journals.

He is a Life Member and was President of The Indian Society of Agricultural Economics, 1989; The Indian Society of Labour Economics, 1995; The Indian Society of Agricultural Marketing, 1999; The Indian Society of Ecological Economics, 2001; and the Indian Economic Association, 2009.

He served as a Member of several expert bodies, including, Seventh and Eighth Finance Commissions; Planning Commission, 1982-1986; Economic Advisory Council to the Prime Minister, 1983-1985; Technical Advisory Committee (TAC) of the Consultative Group on International Agricultural Research (CGIAR), 1981-82 and 1995-1997; Central Board of Directors, Reserve Bank of India, 1994-2000; and the National Advisory Council, 2004-2008.

He also served as Chairman, Steering Committee on Agriculture and Allied Sectors for Formulation of the Eleventh Five Year Plan (2007-2012), Planning Commission, 2007; High Powered Fertilisers Pricing Policy Review Committee, Government of India (GoI), 1997-98; Expert Committee on Rehabilitation and Environmental Aspects of Tehri Hydro-Electric Project, GoI, 1996-97; Advisory Committee, National Accounts Statistics, GoI, 1992-1995; Technical Committee on Drought Prone Area Programme and Desert Development Programme (DPAP & DDP), Ministry of Rural Development, GoI, 1993-94; National Commission on Rural Labour, GoI, 1990-91; Expert Committee for Review of Methodology of Cost of Production of Crops, Ministry of Agriculture, GoI, 1990; and Advisory Council on the Implementation of the 20-Point Programme, GoI, 1987-1989.

Prof Rao was awarded Rafi Ahmed Kidwai Memorial Prize for 1974-75, by the Indian Council of Agricultural Research, for Outstanding Contribution in the field of Agricultural Economics; he is a Fellow, National Academy of Agricultural Sciences, New Delhi; received K.H. Batheja Award for one of the best books published in Economics in India during 1993-94; received the Financial Express Award for Lifetime Work in Economics, in 1995; and was awarded 'Padma Bhushan' by the President of India in 2004 in the field of Literature and Education.

Essays on Inclusive Growth and Some Reminiscences

C.H. Hanumantha Rao

ACADEMIC FOUNDATION
NEW DELHI

www.academicfoundation.com

First published in 2014
by

ACADEMIC FOUNDATION
4772-73 / 23 Bharat Ram Road, (23 Ansari Road),
Darya Ganj, New Delhi - 110 002 (India).
Phones : 23245001 / 02 / 03 / 04.
Fax : +91-11-23245005.
E-mail : books@academicfoundation.com
www.academicfoundation.com

Cataloging in Publication Data--DK
Courtesy: D.K. Agencies (P) Ltd. <docinfo@dkagencies.com>

Hanumantha Rao, C. H.
Essays on inclusive growth and some reminiscences / C.H. Hanumantha Rao.
p. cm.
ISBN 9789332700895

1. Economic development--India. 2. Economic development--India--Andhra Pradesh. 3. Agricultural development projects--India. 4. India--Economic policy. 5. Economists--India. 6. Hanumantha Rao, C. H. I. Title.

DDC 338.954 23

Typeset by Italics India, New Delhi.

Printed and bound by The Book Mint, New Delhi.
www.thebookmint.in

Dedicated to
My Wife, CH. Krishna Kumari

Contents

II
Agriculture: Growth and Sustainability

III
Andhra Pradesh

Introduction

ACHIEVING economic growth with equity or social justice has always been the foremost objective of development planning in India. But the equity goals, e.g., reducing regional disparities in development, employment generation, poverty reduction, etc., were not sufficiently integrated with the growth process itself by according due priority to labour-intensive sectors like agriculture, rural non-farm sector, manufacturing and infrastructure development in the less developed regions. These goals were sought to be achieved largely through special schemes for employment generation, income generating activities targeted for the identified poor, public distribution of food grains at subsidised rates, development of drought-prone areas, etc.

However, being essentially top-down programmes, they suffered from inappropriate designing as well as ineffective implementation. As a consequence, achievements from such schemes fell considerably short of the targets, unlike in the case of GDP growth rates. These deficiencies are ultimately traceable to the prevailing social structure and the nature of our polity which influence the processes of decision-making as well as implementation.

Rising regional disparities in development and slow growth of sectors like agriculture, rural non-farm sector and manufacturing together with inadequate attention to health care and education for over a decade following economic reforms of the early 1990s led to the rise in income inequalities and rural distress. This prompted renewed attention by policy makers to issues of equitable or inclusive growth. The Eleventh Five Year Plan came out with an explicit objective for achieving inclusive growth. Given the backlog of poverty and unemployment, special programmes, particularly for relieving rural distress, needed to be in place, even as efforts are made simultaneously to make the growth process sufficiently inclusive.

The Twelfth Five Year Plan (2012-2017) has clearly spelt out the conceptual framework for achieving inclusive growth. It envisages a growth process that would reduce regional and intra-regional disparities in development, step up the growth of agriculture and manufacturing, give due priority to social sectors like education and health and provide adequate opportunities for the disadvantaged sections of society to participate in the growth process. However, the achievement of these goals is crucially dependent upon the impact of sociopolitical factors, especially at the state level, on the allocation of resources to different sectors, consistent with the objectives of inclusiveness, proper designing and implementation of schemes by improving capabilities at the grass roots level and ensuring accountability through the empowerment of local elected bodies and involvement of civil society at large.

The essays bearing on inclusive growth, included in this book, were written for different occasions over the last decade. They focus on deficiencies in infrastructure—physical, social and institutional—especially in the less developed regions, resulting in the inability to ensure equitable benefits from economic reforms and globalisation; technology fatigue, marketing constraints and sustainable use of water for irrigation in the case of agriculture; and the growth performance in Andhra Pradesh where regional inequalities in the distribution of gains from development gave an impetus to the demand for separate statehood for Telangana. This led to the recent decision of the Government of India to initiate the process for the formation of a separate state of Telangana. The role of sociopolitical factors in accentuating income inequalities in the wake of globalisation and reforms is brought out by bringing in the comparative experience of India and China.

The last section of the book includes some of my reminiscences briefly from my student days since the mid-1940s, to my interactions on economic policy issues concerning equitable development since the 1980s. Many of these reminiscences relate to individuals ranging from veteran freedom fighters in my early days to the eminent scholars and public figures in the government, including some Prime Ministers and distinguished civil servants.

S. Mahendra Dev and E. Revathi read the section on Reminiscences and gave their comments and suggestions which improved the presentation. In the course of the preparation of these essays, I received valuable help from N. Chandrasekhara Rao, C. Ravi, G.R. Reddy and N. Sreedevi. I also received valuable help from B. Sankara Reddy in the Administration, K. Srihari, Computer Engineer and P. Sudhakar, Assistant Librarian. I am grateful to all.

— C.H. Hanumantha Rao
Centre for Economic and Social Studies,
Hyderabad.

I

Growth and Inclusiveness

1

Inclusive Growth

An Overview of Performance and the Challenges Ahead

Global Financial Crisis

Our economy, like many others, has had to suffer the consequences of the worst financial and economic crisis, although we had nothing to do with its origins. Luckily, we have been protected from the severity of its impact primarily because of the prudent management of our financial sector by the Reserve Bank of India (RBI) under the stewardship of the then Governor, Dr Y.V. Reddy, even as the crisis was brewing abroad and in the face of the pressures for more rapid liberalisation of financial markets in the interests of higher growth. Credit also goes to the gradual pace of reforms introduced in the post-reform period and institution-building and policies, including nationalisation of major commercial banks, in the pre-reform period which enabled harmonising of reforms in the financial and external sectors with progress in fiscal and real sectors.

The immediate issue is to address fundamental issues concerning structural changes in the global financial architecture, as recommended by the Commission of Experts of the President of the United Nations General Assembly on Reforms of the International Monetary and Financial System (United Nations, 2009). However, our concerns and efforts notwithstanding, we do not have much influence in this regard and we should be prepared to live with several uncertainties on this front. The options open to us are largely on the domestic policy front to cope with the emerging challenges. The government and the RBI responded promptly with measures to minimise the impact of this crisis on the domestic economy, the government putting in place fiscal stimulus packages designed to tone up domestic demand by strengthening infrastructure and raising purchasing power, particularly in the rural areas. As a result, the economy seems to be recovering steadily, despite the drought hitting agriculture.

From the long-term perspective, the lessons thrown up by the present crisis assume great relevance. First and foremost, market fundamentalism (i.e., the belief that markets are self-correcting, allocate resources efficiently and ensure full employment) has now been called into question as never before after the Great Depression of 1930s because an important factor behind the current global financial crisis is a serious underestimation of the potential for market failures. Second, state failure has been no less glaring. Dr Y.V. Reddy, a member of the Commission of the United Nations, calls it 'Regulatory Capture': Financial markets developed far more rapidly than the real economy and the process fostered considerable linkages with the political economy, that drove the actions and inactions of both the central banks and governments, leading to the failure of both the market and the state—morally and intellectually (Reddy, 2009). Dr Robert A. Johnson, another member of the Commission says, "the incentives faced by public officials, regulators, and elected officials and the role of money in politics are important antidotes to romantic notions of the efficiency of regulation to correct for the market failures. In addition, the role of expertise, and the incentives of experts who themselves are incentivised by considerations of power, prestigious awards and compensation should be thoroughly examined" (Johnson, 2009).

The report of the Commission suggests a third policy lesson for both the developed and developing economies: The rising inequalities in income have led to insufficiency of global aggregate demand exacerbating the present crisis. Hence, the emphasis on inclusive growth and measures for social protection. For large developing economies like India and China, an inclusive growth strategy for raising domestic demand could provide a safety net against the vagaries of global markets. China, which apparently has been hit harder than India because of its greater exposure to globalisation, seems to be already moving in this direction.

The question whether capitalism can survive the present global financial crisis is being debated fiercely. Nearly 70 years ago, in the wake of the Great Depression, Joseph A. Schumpeter in his classic work, *Capitalism, Socialism and Democracy*, had predicted the demise of capitalism and the rise of socialism. But thanks to Keynesianism and its powerful impact on the growth models of both the developed and developing economies, capitalism has survived so far, and socialism is nowhere round the corner.

But socialism and centralised economic planning had a sway for well over half a century, and economic planning had a great appeal in large parts of the globe. However, in a period when technological change and rise in productivity became the predominant source of economic growth in Western capitalist economies, the socialist economies lagged very much behind on account of closed nature of their economies, lack of competition and absence of incentives for innovation and enterprise, which eventually led to the collapse of their systems.

Therefore, the present crisis makes it imperative to trigger growth patterns that could tame both the market and the state through the governance of financial and economic sectors that is inclusive and widely participatory, and so accountable, both in the international and national arena.

Economic Reforms, Growth and Inequalities

The last two decades of globalisation have witnessed rising growth rates in gross domestic product (GDP) along with growing income inequalities in several developing economies including India. The proportion of people living below the poverty line has been declining but at a lower pace than in the pre-reform period, implying that the elasticity of poverty reduction with respect to GDP growth has declined. Further, whereas economic reforms and globalisation have in general led to the rise in GDP growth rates, there is no unambiguous causal relationship between them and rising income inequalities or slow reduction in poverty. This experience suggests that initial or pre-reform domestic socio-economic conditions as well as post-reform policies could be mainly responsible for rising inequalities and slow reduction in poverty (Rao, 2005; Bardhan, 2006; 2009; Kanbur, 2008; Banerjee, 2009). In this presentation, I propose to draw attention to such unfavourable initial conditions and discuss the policies needed for achieving inclusive growth.

Inclusive Growth: Recent Experience in India

Prior to the current global economic meltdown with its adverse impact on the domestic economy, high GDP growth was sustained over the five years with strong recovery of manufacturing sector. This particular configuration brought a boom in tax revenues for the government, led by the revenues from direct taxes, especially from the corporate sector. This enabled governments, both at the Centre and the states, to step up public investment in infrastructure and other expenditures on agriculture and social sectors, demonstrating that for a resource constrained economy, high GDP growth is a necessary, though not sufficient, condition for achieving inclusive growth (Rao, 2009; Ahluwalia, 2009).

Public investment in agriculture which had declined from 5 per cent of agriculture GDP in the early 1980s to below 2 per cent in 2002-03 has now been stepped up to over 3 per cent, with the target of raising it to 4 per cent by the close of the Eleventh Plan. Together with private investment, the overall capital formation in agriculture amounted to 16.1 per cent of agriculture GDP in 2007-08 (Government of India, 2009a), which is the highest in the last 25 years (Planning Commission, 2008a). The target of doubling the supply of institutional credit to agriculture in 3 years was exceeded. There have been concerted policy interventions such as launching of the National Food Security Mission and the Rashtriya Krishi Vikas Yojana.

As a result, there is a clear upturn in agriculture, the growth rate of agriculture GDP rising to over 4 per cent during the five years ending 2007-08, with comfortable stocks of food grains before the onset of current drought—almost comparable to that achieved when the green revolution was in full swing.

National Rural Employment Guarantee Scheme, recently renamed as Mahatma Gandhi National Rural Employment Guarantee Scheme (MGNREGS), has been extended throughout the country. Most evaluations—official as well as non-official—show that the implementation of this scheme has been far more effective than any of its predecessor schemes in the country. In particular, the leakages have been reduced significantly in many places in respect of the number of days a worker is employed as well as wages paid. This is attributed mainly to the adoption of innovative methods like opening post office and bank accounts for payment of wages to workers and the use of information technology, apart from social auditing. The significant rise in the rural market wages reported from several parts of the country and the widespread demand from the farmers to reschedule the operation of the employment guarantee scheme from busy to lean season testify to the effectiveness of income generation for the poor from its implementation.

Coming against the backdrop of rural distress for over a decade, the scheme was successful in providing much needed relief to the rural poor, apart from raising their awareness and bargaining power. Further, the recent experience shows that the Right to Information Act being implemented in the country holds a great promise for improving the effectiveness of this and similar schemes designed to improve the lot of the poor.

Major initiatives have also been taken in the social sectors like education and health through Sarva Siksha Abhyan (SSA) and National Rural Health Mission (NRHM). The allocations for these social sectors have been stepped up substantially in the 11th Plan when compared to the 10th Plan. The allocation of funds to the poorer states from the Centre during 2007-08 for programmes like NRHM, SSA and poverty alleviation was well above their share in population.

Some other notable initiatives in this field are strengthening the Mid-Day Meal Scheme, the Right to Education Act and the National Food Security Act. Government has also announced the Programme for the Welfare of Minorities aimed at enhancing opportunities for education, employment, credit support and ensuring an appropriate share in infrastructure development schemes. It envisages location of certain proportion of development projects in minority concentration areas; and provides that, wherever possible, 15 per cent of targets and outlays under various schemes should be earmarked for minorities

Acting on the recommendations on Social Security made by the National Commission for Enterprises in the Unorganised Sector

(NCEUS), headed by Arjun K. Sengupta (NCEUS, 2006), government has introduced schemes to provide Social Security Coverage through life cover, health insurance and extension of old age pension on the lines recommended by the NCEUS, but by restricting coverage to below the poverty line (BPL) households. With the coming into force of the Unorganised Workers Social Security Act, 2008, over 90 per cent of the unorganised workers in the country would get the benefit of health, life and disability insurance, old-age pension and group-accident scheme.

These initiatives taken in the recent period for achieving growth with equity are significant. Even so, considering the magnitude of challenges, there is a compelling need to mount a major effort to ease the infrastructural constraints in the economy, especially in the less developed areas, and, equally important, evolve a more comprehensive and active social policy to facilitate inclusive growth as well as to cope with inequitable distribution of gains from growth on account of the prevailing iniquitous social structure.

Inclusive Growth: The Challenges Ahead

Deficiencies in Infrastructure: Regional and Rural-Urban Divides

Regional Disparities

The Indian economy continues to be constrained primarily by the insufficiency of physical and social infrastructure where the government plays a major role [Economic Advisory Council (EAC), 2009]. At a little over 4 per cent of GDP in the base year of the Eleventh Plan, public investment in infrastructure was seriously deficient. This is explained by years of underinvestment relative to the requirements (World Bank, 2008). Inadequacy of infrastructure in the less developed regions and rural areas in general in the country is responsible for the inability of such areas to fully benefit from the opportunities opened up by economic reforms including globalisation, leading to growing regional and rural-urban disparities.

Generation of adequate employment opportunities, particularly for the socioeconomically disadvantaged sections, is key to achieving inclusive growth. In this context, the deceleration of employment growth—more so for weaker sections—in the post-reform decade ending 2004-05, when there was an acceleration in GDP growth, should be a matter of concern (NCEUS, 2009; Papola, 2009). Broad-based or balanced growth between different regions and between rural and urban areas would generate greater aggregate employment from a given rate of investment than unbalanced growth because capital- labour ratios are generally low in the less-developed and low-wage areas. Higher rate of investment for infrastructure development in such areas, therefore, becomes critical for greater employment generation.

In the present context of globalisation, associated with trade and financial liberalisation, the contribution of information and communication technology to GDP growth is rising sharply and becoming the driving force behind rising inequalities, as it raises the demand for skills that are in short supply and highly unequally distributed between different regions, income groups and between males and females (Bhattacharya, 2008; Nauriyal *et al.*, 2009). This underlines the need for an effective domestic policy to expand the opportunities for education and acquisition of necessary skills in the less developed areas and for the socioeconomically disadvantaged sections.

According to the Eleventh Plan, higher level of infrastructure development is envisaged "especially in rural areas, and in the remote and backward parts of the country consistent with the requirements of inclusive growth at 9 per cent per year" (Planning Commission, 2008a). But from the information given in the Plan document, it is not possible to find out how the infrastructure planned is going to be spread over different states, on the one hand, and between agriculture and the non-farm sector including small towns, on the other.

For the Eleventh Plan, the targeted state-wise growth rates in respect of gross state domestic product (GSDP) do show smaller divergence between the richer and poorer states than those achieved in the Tenth Plan. But the absence of targets in respect of investment rates castes doubt on the credibility of these growth targets. For one thing, the per capita plan outlays (including central assistance) of the poorer states have been declining relative to those of the richer states in the post-reform period. The poorer states have been handicapped basically by their own weaker resource position. Because of lower tax-GSDP ratios and higher debt-GSDP ratios, their credit-worthiness is lower, and so they have not been able to access borrowings adequately from the market. Private investment has been flowing basically to the high income states where infrastructure is well developed (Rao, 2006).

The position has now become worse for the poorer states because of the Centre's decision to do away with the loan component of the normal central assistance for state plans. No alternative mechanism has been put in place so far to enable the poorer sates to easily access loans from the market. There is no increase in central assistance in the Eleventh Plan for infrastructure development of the backward regions through Backward Regions Grant Fund, Special Plan for Bihar and Action Plan for the undivided KBK (Kalahandi-Bolangir-Koraput districts of Orissa), which have only been protected at the Tenth Plan level. Such assistance constitutes less than 1 per cent of total Eleventh Plan outlay for all states. It is not known how the performance of Bharat Nirman—a flagship programme designed to create infrastructure—is tilted in favour of the less developed states.

In view of the steep decline in the share of public investment in the overall investment, the Planning Commission admits that there is "a very great limitation on the influence that fiscal quantities, allocations and strategy can directly exert on growth rates, especially at state level", and suggests that "states have, therefore, to focus on providing the necessary policy framework and supporting environment that makes economic activity possible and attractive enough for private sector investments" (Planning Commission, 2008a).

But, can such a 'supporting environment' make good the gaps in public investment in infrastructure? Historically, due to the differences in land tenure systems and other factors, the developed states have had relatively more efficient systems of governance in terms of skills, responsiveness and the quality of delivery systems (World Bank, 2006). It should be noted in this context that unlike capital—which is highly mobile across regions and continents—good governance cannot be transplanted in an area, as it evolves basically within the prevailing sociopolitical structure over a long period.

There is, however, a silver lining in the experience of recently created smaller states, *viz.*, Chattisgarh, Jharkhand and Uttaranchal, showing that these states achieved growth rates far exceeding the targets set for the Tenth Plan period whereas the performance of their parent states, *viz.*, Madhya Pradesh, Bihar and Uttar Pradesh, fell considerably short of the targets (Planning Commision, 2008). Of late, Bihar too has been growing faster because, among other things, it has become smaller and viable following the creation of Jharkhand. Apart from releasing the creative energies of the people, viability of smaller states may have contributed to better governance, attracting greater private investment from outside as well as planning and utilising resources more efficiently (World Bank, 2007).

Agriculture

The imbalance between the global supply and demand for food has been building up over a period of two decades owing to a steady decline in investment in agriculture in several developing economies and decline in global public expenditures on agricultural research through the Consultative Group on International Agricultural Research (CGIAR) designed to benefit the resource-poor farmers in the developing economies. Consequently, the divergence between the growth rate of overall GDP and that from agriculture has been increasing, particularly in the fast-growing economies like China and India. This explains, on the one hand, rising demand for food, especially for milk, meat, vegetables and fruits, and, on the other, slow growth in their supplies.

The large-scale diversion of cereals to ethanol production in the developed countries has aggravated the demand-supply imbalance for food to crisis proportions. Thus the nature of demand for food grains could be fundamentally altered from inelastic, or one responding only slowly to rise in incomes, to highly elastic on par with that for energy. These developments threaten to reverse the historical trend of a decline in the relative prices of food grains, adversely impacting the lives of the poor (Rao, 2008).

Agricultural strategy in India has to be robust enough to face such an adverse global food scenario. Apart from stepping up public investment in irrigation on which, as mentioned above, some progress has been made in the last few years, other major challenges in agriculture are: rejuvenating dry land areas through soil and water conservation; revamping of agricultural research and extension; developing the marketing infrastructure, including transport and cold storage, to meet the growing needs of diversified and high-value agriculture by linking the groups of farmers to processors and organised retailers (Gulati, 2009a; 2009b); and reviving and strengthening of various public support systems for small and marginal farmers.

The Eleventh Plan strategy for soil and moisture conservation rests on watershed development in rainfed areas under the aegis of the National Rainfed Areas Authority. This superstructure can yield the desired results only if the focus on participatory approach at the grassroots level is revived.

Green Revolution in the 1960s was triggered off mainly by the availability of yield-increasing technology. There is no comparable breakthrough in agricultural technology in the recent period except in the case of cotton, supported by the private sector. Agricultural research in the public domain in India has suffered on account of paucity of resources—about 0.7 per cent of agriculture GDP compared to over 1.2 per cent in China—and excessive centralisation and bureaucratisation. State agricultural universities (SAUs), which are responsive to location-specific needs, need to be allocated considerably more resources with high priority to modern biotechnologies suited to dry land areas and small farms. Agricultural extension, which has been in disarray for quite some time, has to be revived for making knowledge accessible to small farmers—mostly women—who now predominate farming in India.

The decline in the relative contribution of agriculture to GDP is leading to out-migration of male labour resulting in feminisation of agriculture (Agarwal, 2004). While this is adding to the work burden on women, it is also opening up new and more remunerative opportunities for them such as dairying, horticulture, fisheries and agro-processing in which they have been traditionally engaged. These activities being land-saving, the constraints faced by women farmers lacking in titles to land could be eased to some extent provided they can get an easy access to the needed resources and services. This would be possible only if women officers and functionaries are inducted in large numbers in the services chain for extension, credit delivery, marketing, and insurance, etc. (Krishnaraj, 2007).

The impact of the global financial crisis on our economy has been compounded by the severe drought hitting several parts of the country this year, leading to the sharp rise in the prices of food grains. It has revived the familiar question as to what extent our efforts towards drought-proofing of agriculture have been successful. It also raises a new question about the impact of global warming on the level of rainfall in India, its distribution, variability and predictability, on which we need to improve our understanding. On the first question, there are indications that the variability in agricultural output has declined perceptibly in the recent period under the combined impact of increase in irrigation, improvements in technology and diversification of agriculture in dry land areas (Chand and Raju, 2009; Planning Commission, 2012). However, even after the available irrigation potential is fully exploited, over half the cultivated area would still be dependent on rainfall directly, or indirectly through minor irrigation. Moreover, because of the steep decline in the average size of holding and soil degradation the vulnerability of small and marginal farmers to droughts would increase. Therefore, even if there is some further drought-proofing of agriculture, drought-proofing of the vast majority of livelihoods dependent on agriculture can be achieved only with a major augmentation of their earnings from non-agricultural occupations.

Rural Non-Farm Sector

For this to happen, however, there is no escape from high growth of agriculture. The studies on the growth of rural non-farm sector in India show that although output and employment have been growing in this sector, it has not really taken-off so far because of slow growth of agriculture, insufficient development of rural infrastructure and weak public support systems. Therefore, along with high agricultural growth, there is a need for providing an enabling environment through augmentation of rural infrastructure like roads, power and communications, increasing the institutional credit flow to the small and tiny sector, building of strong public support systems for improving market information and marketing networks, skill and technology upgradation and their dissemination through training, especially for the weaker sections, and for promoting quality consciousness and quality control (Nayyar and Sharma, 2005; Papola and Sharma, 2005; Chadha, 2009).

Social Structure and Social Policy

Conventional analyses of poverty and inequality in India seldom throw light on the processes of social and economic discrimination by which poverty and destitution are perpetuated in a hierarchical society based on group identities such as gender, caste, class, ethnicity and religion.

The prevailing social structure is characterised by inequalities in land holdings and other forms of wealth, status and power. Because of this, achieving inclusive growth is far more challenging than stepping up the GDP growth rate *per se*. This is borne out by the fact that, in the past, slippages in realising targets in respect of sectors bearing on inclusive growth, e.g. agriculture, poverty reduction, primary health, elementary education and reducing regional disparities in development, etc., have been far more significant than the slippages in respect of GDP growth.

Even more serious, the population belonging to socially and economically disadvantaged sections like scheduled castes (SCs), scheduled tribes (STs) and backward classes, large majority of minorities, and women and children have benefited least from growth and rising prosperity. The slow improvement in the various development indices for these sections and the slow rise in public expenditure targeted to them as well as for social sectors, like health, education and poverty alleviation programmes, and greater vulnerability of the poor to various kinds of risks in this period have been documented extensively in the literature (Radhakrishna and Ray, 2005; Government of India, 2006; Dev, 2008; Kannan, 2009).

Growth, by itself, far from reducing the risk of poverty for the socially excluded groups has in fact increased such risk due to market failures arising from discrimination as well as public interventions that are both inadequate and discriminatory in operation. A two-pronged strategy to address the problem is called for, first, 'social and economic empowerment' of the excluded groups by improving ownership of capital assets like land and business, and education and skills and, second, providing 'equal opportunity' for them through legal safeguards, reservations and similar measures in public as well as private sectors and by ensuring adequate participation in governance at various levels. Achieving equity through such measures will not only restore self-esteem and confidence among the socioeconomically disadvantaged sections, but also improve the overall efficiency of the economic system and ensure the sustainability of growth momentum by tapping the vast underutilised human potential (Thorat and Newman, 2009).

Apart from social security for the unorganised sections of labour on which, as mentioned above, a good beginning has been made, a comprehensive social policy at this stage should have four major elements, *viz.*, land policy, rehabilitation and resettlement policy, financial inclusion and inclusive governance.

Land Policy

While the policy of 'Land to the tiller' which was the overarching principle in the early post-independence period continues to be relevant, especially for tribal areas, half a century of development has brought some new concerns to the forefront. With growing industrialisation and urbanisation, the rising demand for land for industrial purposes, including Special Economic Zones, and for housing, transfer of land from tillers to the emerging non-farm activities, is becoming inevitable. While raising agricultural productivity is a must to cope with the shrinkage of agricultural land, the slow growth of non-farm opportunities for employment and livelihoods and lack of social security for small holders argue for a careful and calibrated approach for land acquisition.

Land alienation continues to be a serious problem in tribal areas. The enactment of Scheduled Tribes and Other Traditional Forest Dwellers (Recognition of Forest Rights) Act, 2006, is a major step towards protecting the rights of STs and other traditional forest dwellers. However, past experience suggests that the implementation of this Act is going to be extremely challenging. A pre-requisite for effective implementation is improvement in land administration in tribal areas, greater say for tribals in the implementation process, and effective monitoring and review, backed by necessary political will to counter the pressures from the vested interests. But the efforts on the ground so far do not seem to match the challenges involved (Planning Commission, 2008b; Government of India, 2009b). According to Shri Digvijay Singh, former Chief Minister of Madhya Pradesh, who has a long experience of dealing with problems of tribal development, "unless we change the laws that focus on exploiting the natural resources at the expense of human beings and the feeling of alienation among tribal population is removed through pro-people steps, we may end up waging a war against our own people" (Singh, 2009).

Policy for Rehabilitation and Resettlement

Willingness to part with land by the farmers for development projects has to be the guiding principle for making the sale transaction just and humane. In the present context of rising land values and cost of living with increasing uncertainty, traditional norms for compensation and rehabilitation and resettlement of displaced persons are proving to be unviable. Government and industry should, therefore, supplement the principle of fixed 'compensation', even if fair at a point of time, with measures for assuring original land owners or the displaced persons adequate stake, on a secure footing, in the new projects and enterprises including those in Special Economic Zones (SEZs) (Singh, 2009). Such measures, adjudicated by independent bodies of experts, would be fully justified by the high and rising profitability of new ventures.

Financial Inclusion

There was a rapid rise in the saving rate in the economy following the nationalisation of banks which forced them to open their branches in the remote areas of the country. This rise in saving rate provided an impetus for the spurt in the growth rate of the economy after the mid-seventies (Basu, 2008). The gross domestic saving from the household sector as per cent of GDP at current market prices rose by 10 percentage points in the pre-reform period itself, following bank nationalisation, from around 9 per cent of GDP in 1968-69 to 19 per cent in 1990-91, as against a rise by mere 2.5 percentage points between 1950-51 and 1968-69 (Government of India, *Economic Survey 2006-07*).

Although following bank nationalisation, there was a phenomenal expansion in the banking services in rural areas and the supply of bank credit to farmers increased significantly, there was a severe retrogression in these respects for quite some time in the post-reform period (Rao, 2005). The United Progressive Alliance Government after coming to power in 2004 tried to provide a corrective by doubling the supply of bank credit to agriculture in three years. But no sub-targets were set for the supply of credit to the small and marginal farmers. The Steering Committee on Agriculture and Allied Sectors for the Eleventh Five Year Plan noted that "the share of direct accounts with a credit limit of less than ₹ 25,000 in total direct accounts declined from 97 per cent in 1990 to 67 per cent in 2005, while their share in outstanding direct credit declined from 0.66 per cent to 0.23 per cent in the same period. Spatial distribution of credit across states and within states continues to remain unequal. The credit disbursed by commercial banks covers only 11.7 per cent of operational holdings in the country" (Planning Commission, 2007; Reddy, 2007).

The share of small-scale industries in total bank credit declined from 12.5 per cent in 1991 to 5 per cent in 2003: the small and marginal farmers account for only about one-third of institutional credit for agriculture, the remaining two-thirds going to the farmers above 2 hectares (Shetty, 2009).

Financial inclusion could no doubt be inhibited by the higher transaction costs of dealing with a large number of small accounts rather than a small number of large accounts. But such costs can be reduced through organisational innovations or, where necessary, met through explicit subsidies to the banks and other institutions. Subsidies can also be extended to make up the losses on account of lower rates of interest charged to the weaker sections. In practice, however, restriction of bank credit to such sections is inhibited not so much by higher transaction costs or risks but by the financial sector and banking sector reforms implemented in the 1990s which have also resulted in the emergence of an inequitable interest rate structure. For instance, paradoxically, a small farmer was made to pay an interest rate of 12 per cent while a highly rated corporate entity could raise money from banks at 6 per cent (Mujumdar, 2009).

The continued neglect of the weaker sections in the provision of bank credit is attributable to the prevailing mindset of the banking system (Rao, 2003). The RBI seems to be aware of the need to change this mindset (Reddy, 2004), but, obviously, the process of change is not going to be easy and demands strong commitment as well as ingenuity in devising appropriate delivery systems.

Inclusive Governance

Governance is not simply a matter of particular administrative arrangements, procedures, monitoring, etc. In view of the all-pervading influence of the prevailing social structure on decision-making as well as implementation processes, governance has to be shaped in the context of on-going social change through the functioning of our democratic system. It is wrong to attribute the existing deficiencies in decision-making and tardiness in the implementation to the prevailing democratic system as such. Rather, they can be traced to the way the functioning of our democracy has been twisted and moulded by the pulls and pressures from the vested interests. Unlike in mature western democracies where democratic institutions evolved in the course of radical social transformation over a long period, in India, institutions of democracy were adopted even as the traditional and iniquitous social structure remained largely intact. In our situation, therefore, democracy and state power have to become instruments for social transformation. This requires broadening and deepening of our democracy by effectively extending it to the grass roots.

Experience has amply demonstrated that participatory or inclusive governance is indispensable for achieving inclusive growth. This is woefully lacking at the grass root levels where the formulation of schemes is not quite in keeping with the local needs and circumstances and administration is not accountable to the people. It is precisely to cope with such challenges that *panchayati raj* was visualised by the late Prime Minister Shri Rajiv Gandhi.

Thanks to the 73rd and 74th amendments to the Constitution, elected *panchayats* have been in place. But they are deprived of necessary functions, finances and functionaries. No serious effort has been made so far to devolve such powers, not to speak of any effort to build up local capabilities by training the elected functionaries. On the other hand, certain state governments have been active in pursuing top-down development and governance by floating numerous schemes and parallel implementation structures which cut at the very root of panchayati raj institutions (PRIs)—on the top of it by naming such schemes after Rajiv Gandhi! This is explained solely by the resistance from the entrenched interests to part with their powers. Therefore, the change in attitudes has necessarily to be driven by the political process.

It is naïve to expect a genuinely inclusive governance without inclusive politics or empowerment of social groups hitherto not having access to political power at different levels. The recent decision of the Central Government to enhance reservation for women in local bodies to 50 per cent and the on-going efforts to provide 33 per cent reservation for women in the Parliament and state legislative bodies, when successful, should go a long way towards ensuring inclusive growth through inclusive politics and governance.

References

Agarwal, Bina (2004). *Gender Inequalities: Neglected Dimensions and Hidden Facets*. 2002 Malcolm Adiseshiah Memorial Lecture. Chennai: Madras Institute of Development Studies.

Ahluwalia, Montek S. (2009). "Growth, Distribution and Inclusiveness: Reflections on India's Experience", in Kaushik Basu and Ravi Kanbur (eds.), *Arguments for a Better World: Essays in Honour of Amartya Sen*. Vol.II: Society, Institutions and Development. New Delhi: Oxford University Press.

Banerjee, Asis (2009). *Globalization and Inequality: A Review of Evidence.* Keynote Paper presented at the National Seminar on Globalization and Inequality. Bengal Economic Association in collaboration with the Indian Economic Association & CSESD, Rabindra Bharati University, Kolkata. June 20-21.

Bardhan, Pranab (2006). "Globalization and Limits to Poverty Alleviation", in Pranab Bardhan, Samuel Bowles and Michael Wallerstein (eds.), *Globalization and Egalitarian Distribution.* New Delhi: Oxford University Press.

———. (2009). "Economic Reforms, Poverty and Inequality in India and China", in Kaushik Basu and Ravi Kanbur (eds.), *Arguments for a better World: Essays in Honour of Amartya Sen,* Vol. II: Society, Institutions and Development. New Delhi: Oxford University Press.

Basu, Kaushik (2008). "India's Dilemmas: The Political Economy of Policymaking in a Globalised World", *Economic & Political Weekly* XLIII (5), February 2-8.

Bhattacharya, B.B. (2008). "Education, Skill formation and India's Economic Development", *The Indian Economic Journal* 56(1): 3-10.

Chadha, G.K. (2009). "Agriculture and Rural Industrialisation in India: Recent Developments and Future Choices", in S. Mahendra Dev and N. Chandrasekhara Rao (eds.), *India: Perspectives on Equitable Development.* New Delhi: Academic Foundation.

Chand, Ramesh and S.S. Raju (2009). "Instability in Indian Agriculture During Different Phases of Technology and Policy", *Indian Journal of Agricultural Economics* 64(2): 187-207.

Dev, Mahendra, S. (2008). *Inclusive Growth in India: Agriculture, Poverty, and Human Development.* New Delhi: Oxford University Press.

Economic Advisory Council to the Prime Minister (2009). *Economic Outlook for 2009/10.* New Delhi. October.

Government of India (2006). *Report of the High-Level Committee on the Social, Economic, and Educational Status of the Muslim Community of India.* New Delhi: Government of India.

———. (2007a). *Report of the Committee on Financial Inclusion.* New Delhi: Ministry of Finance.

———. (2007b). *Economy Survey 2006-2007.* New Delhi: Ministry of Finance.

———. (2009a). *Economic Survey 2008-09.* New Delhi: Ministry of Finance.

———. (2009b). *Report of the Committee on State Agrarian Relations and the Unfinished Task in Land Reforms.* New Delhi: Department of Land Resources, Ministry of Rural Development.

Gulati, Ashok (2009a). *Emerging Trends in Indian Agriculture—What can we learn from these*?, 2nd Prof. Dayanatha Jha Memorial Lecture. New Delhi: National Centre for Agricultural Economics and Policy Research, ICAR.

———. (2009b). "How to Balance Food Import & Farmer Incentives—Focus on Incentives, Imports and Domestic Production", *Economic Times.* New Delhi. November 26.

Johnson, Robert A. (2009). "Failure of an Economic Philosophy and Implications for Financial Regulation", *Economic and Political Weekly* XLIV(36): 31-33. September 5-11.

Kanbur, Ravi (2008). *Poverty and Distribution: Twenty Years Ago and Now,* Fifth Lecture in Golden Jubilee Series. New Delhi: Institute of Economic Growth, University of Delhi.

Kannan, K.P. (2009). "Dualism, Informality and Social Inequality: An informal Economy Perspective of the Challenge of Inclusive Development in India", *The Indian Journal of Labour Economics* 52(1): 1-32. January-March.

Krishnaraj, Maithreyi (ed.) (2007). *Gender, Food Security and Rural Livelihoods.* Chennai: M.S. Swaminathan Research Foundation. Published by STREE, Kolkata.

Majumdar, N.A. (2009). *Journey Towards Holistic Development: From Milton Friedman to Mahatma Gandhi*. Seventh Prof. G.S. Diwan Memorial Lecture. Mumbai: DS Actuarial Foundation Services.

National Commission for Enterprises in the Unorganised Sector (2006). *Report on Social Security for Unorganised Workers*. New Delhi.

———. (2009). *The Challenge of Employment in India: An Informal Economy Perspective*, Vol.I—Main Report. New Delhi.

Nauriyal, D.K., Bimal K. Sahoo and Anil Dixit (2009). "Economic Growth, Globalisation and Human Capital: Empirical Evidence from India", *The Indian Economic Journal* 56(4): 37-54, January-March.

Nayyar, Rohini and Alakh N. Sharma (2005). "Introduction and Overview", in Rohini Nayyar and Alakh N. Sharma (eds.), *Rural Transformation in India: The Role of Non-Farm Sector*. New Delhi: Institute for Human Development.

Papola, T.S. (2009). "India: Growing Fast, But also Needs to Industrialise!", *The Indian Journal of Labour Economics* 52(1): 57-70, January-March.

Papola, T.S. and Alakh N. Sharma (2005). "Towards a Policy Agenda for Rural Non-Farm Sector", in Rohini Nayyar and Alakh N. Sharma (eds.), *Rural Transformation in India: The Role of Non-Farm Sector*. New Delhi: Institute for Human Development.

Planning Commission (2007). *Report of the Steering Committee on Agriculture and the Allied Sectors for the Formulation of the Eleventh Five Year Plan (2007-2012)*. New Delhi: Government of India.

———. (2008a). *Eleventh Five Year Plan (2007-2012)*, Vols. I and III. New Delhi: Government of India.

———. (2008b). *Development Challenges in Extremist Affected Areas—Report of an Expert Group to Planning Commission, Vols. 1 and 2*. New Delhi: Government of India.

———. (2012). *Twelfth Five Year Plan (2012-2017)*. New Delhi: Government of India.

Radhakrishna, R. and Shovan Ray (eds.) (2005). *Handbook of Poverty in India—Perspectives, Policies, and Programmes*. New Delhi: Oxford University Press.

Rao, C.H.H. (2003). "Reform Agenda for Agriculture", *Economic & Political Weekly* 38(7). February 15.

———. (2005). *Agriculture, Food Security, Poverty, and Environment: Essays on Post-reform India*. New Delhi: Oxford University Press.

———. (2006). "Growing Regional Disparities in Development in India: Post-Reform Experience and Challenges Ahead", Lecture dedicated to the Memory of Professor A.M. Khusro, 88th Annual Conference of the Indian Economic Association, Visakhapatnam. December 27-29, 2005; in *The Indian Economic Journal* 54(1): 52-61. April-June.

———. (2008). "Survey of Indian Agriculture", *The Hindu*.

———. (2009). "Inclusive Growth: Recent Experience and Challenges Ahead", *Economic and Political Weekly* XLIV(13): 16-21. March 28-April 3.

Reddy, G.R. (2007). *Agricultural Credit in India: Recent Trends* (mimeographed). Hyderabad: Centre for Economic and Social Studies.

Reddy, Y.V. (2004). "Rural Credit: Status and Agenda", *Reserve Bank of India Bulletin*, December.

———. (2009). *India and the Global Financial Crisis: Managing Money and Finance*. New Delhi: Orient BlackSwan.

Shetty, S.L. (2009). *Growing Inequality: A Serious Challenge to the Indian Society and Polity*, Dr V.K.R.V. Rao Memorial Lecture-6. Bangalore: Institute for Social and Economic Change.

Singh, Digvijay (2009). "How to Combat the Maoists", *The Economic Times*, New Delhi, October 23.

Thorat, Sukhadeo and Katherine Newman (eds.). (2009). *Blocked by Caste: Economic Discrimination in Modern India*. New Delhi: Oxford University Press.

United Nations (2009). *Report of the Commission of Experts of the President of the United Nations General Assembly on Reforms of the International Monetary Financial System*. New York.

World Bank (2006). *Inclusive Growth and Service Delivery: Building on India's Success*, World Bank Development Policy Review. Macmillan India Ltd.

———. (2007). *Jharkhand: Addressing the Challenges of Inclusive Development*. Rural Poverty and Economic Management, India Country Management Unit, South Asia.

———. (2008). *Report of The Commission on Growth and Development*, Washington.

This chapter is a revised version of the Presidential Address at the 92nd Annual Conference of the Indian Economic Association (27-29 December, 2009), School of Management, KIIT University, Bhubaneswar; originally published in *The Indian Economic Journal* 58(1), April-June, 2010.

2

Economic Reforms with Human Face

IN 1991, a World Development Report dealing with 'State and Market' was released by the World Bank. This was considered as the most important issue in development at that time. After 14 years, it continues to be the most important issue in development. The major difference in India is that economic reforms introduced in 1991 are sought to be implemented now with a Human Face. This change in approach has something to do with the experience of the last 14 years, which prompts us to implement economic reforms with a special emphasis on social equity.

Fourteen years after the introduction of economic reforms in the country is a long enough period to justify its evaluation. This experience has thrown up a mixed picture on economic reforms. We cannot assert that with the introduction of economic reforms, the country is decidedly on a higher growth path now. We cannot also say that the economy has been disrupted and poverty has increased.

On the positive side, perhaps the most important achievement is in the external sector. The economic reforms in 1991 were triggered off by the balance of payments crisis. Basically the objective at that time was to address this issue and the most important achievement is in that very same sector. We have very comfortable foreign exchange reserves—in fact, embarrassingly large reserves. We have not been able to put them to the most effective use for development purposes. But this is, nevertheless, a significant achievement considering the state of affairs at that time. Next, we have been able to maintain a reasonably high rate of growth in gross domestic product (GDP). In fact, in the post-economic reform period, the rate of growth has been slightly higher than in the earlier period. Certainly, in per capita terms, it is higher than before, although in respect of the commodity sectors, e.g., agriculture and industry, the rate of growth has shown a significant decline in the post-economic reform period. It is basically due to the improvement in the growth of the services sector that we have been able to accelerate the GDP growth. These are clearly two achievements.

Contrary to earlier expectations, the overall growth potential in the next few years is not encouraging because the rise in the savings and investment rate is much below the expectations. So whether we will be able to sustain this high rate of growth in the immediate future is in question because the experience of other better performing countries has been a significant rise in the savings and investment rates. There is in fact a decline of savings in the public sector in our country. That is really the most disquieting factor, which has led to a decline in public investment.

Along with the rise in the rate of growth in GDP, there has been some decline in poverty. There is a controversy in regard to poverty decline. Basically the controversy is about the rate of decline. Some say that the rate of decline has been higher in the post-reform period than in the earlier period. Some other scholars have suggested that poverty decline has slowed down. But there is a consensus that there has been a decline in poverty.

Corresponding to the fall in public savings, we find that the fiscal deficit has increased in the recent period. The position is as worrisome as in the early 1990s when economic reforms were introduced. In fact, the combined fiscal deficit of the Centre and states may be even higher than in 1991, although it does not spill over to the external sector as is usually the case. But today despite our comfortable position in the external sector, rising fiscal deficit is a very disquieting aspect of economic scene.

On the negative side, the most worrying aspect is the decline in the employment growth. If we look at the picture of 1980s and now, the rate of growth in employment, particularly rural employment growth, has decelerated very much. This is because the growth in GDP has not been of a nature which generates greater employment. It is not an employment-generating type of growth. That is to say, the rate of growth of employment in relation to the growth of GDP or the employment elasticity has come down drastically. In fact, in some sectors, like the primary sector, employment elasticity has become virtually zero.

What are the reasons for these negative results? Generally, we ask whether the content of economic reforms has been good in terms of, say, right priorities, i.e., whether agriculture has been given due priority and whether infrastructure development has received adequate attention. Then there is an issue of sequencing of reforms. Some countries have started with reforms in agriculture and achieved very good results in terms of reducing poverty and improving the well-being of the rural population. So the issue is whether reforms have moved far ahead on certain fronts by slowing down very much in certain very desirable areas. The third issue, which is often discussed in the context of reforms, is the speed with which reforms are implemented. In 1991, the World Bank report had suggested that one should proceed with the maximum speed in implementing reforms. First of all, if reforms were to proceed on all fronts simultaneously, it means that there is no question of priority and all the sectors are equally important.

What was really surprising was the suggestion that, particularly when the governments are vulnerable and the political situation is fluid, that is the right moment for speeding up reforms. But this is precisely how reforms can become counter-productive. If we are to implement economic reforms with highest possible speed when the political situation is very fluid or even the governments lack credibility, then such a course of action would discredit governments even further. The consequences would be such that it would be difficult for the governments to sustain economic reforms. So we have to have a proper system of priorities and proper sequencing, and the speed which is affordable or absorbable by the given system in the country. We have to learn from doing and build upon our experience and then proceed further.

However, so far as speed is concerned, in our country, we do not run any risk of proceeding very fast. In some other countries like China with a different political system that can implement programmes with a great speed, they can achieve good results if their priorities are right but can also lead to disastrous consequences, if priorities are not right. Therefore, we should work for a system where we can take right decisions and implement them properly.

In order to frame a right system of priorities and work for a right manner of implementation, it is good to look at the international experience in economic reforms from where we can learn a great deal. One can classify these cross-country experiences into five broad categories. One is the Latin American and African experience, which came in the 1980s, out of the debt crisis where the results have been very disastrous. To this day, people remember the prescriptions of the World Bank and International Monetary Fund (IMF). There was a lot of criticism of their prescriptions because of that experience. This is a situation where the distribution of wealth and income is highly unequal as compared to any other part of the world. That is to say, the initial conditions there were not favourable both in terms of the distribution of income and wealth as well as the institutional framework and the governance typified by the very short tenure of many of the authoritarian regimes in these countries. Further, there was a very serious debt crisis, which also constituted a very unfavourable initial condition.

The best case is that of the East Asian countries, that is, China, Korea and Taiwan where economic reforms have yielded spectacular results. The initial conditions there were very favourable. Land reforms were successful much before economic reforms were introduced. Peoples' Republic of China and Korea are the two diametrically different socioeconomic systems in respect of ideologies, but in both the places, land reforms were successful. And, the level of social development e.g., health care system, literacy levels, primary education system, was very high and demographic transition was almost complete at the time when reforms were introduced. What is most important, their governance was such that they did not borrow the prescriptions mechanically from the international financial institutions. In fact, the reforms came not as prescriptions from above but were indigenously designed. So, unlike the Latin American situation, these countries had full freedom to design reforms in a way that suits their own conditions.

The worst case is that of transitional economies—the former Soviet Union, Central Asian countries and some of the other former socialist countries. Economic reforms there led to instability and unrest as in the former Soviet Union or the Central Asian countries. The consensus now is that due to the lack of institutional preparedness for major economic reforms there, the classical World Bank package coming from above led to a very difficult situation.

On the favourable side again, there is the case of South-East Asian countries whether it is Indonesia, Philippines or Thailand. Although their social structure was not as favorable as in East Asian countries, the initial conditions were reasonably favourable in terms of infrastructure development, levels of social development and land reforms.

The fifth case is our area, that is, South Asian countries—India, Nepal, Bangladesh, Sri Lanka and Pakistan. We have pointed out two very successful cases and two very bad cases. South Asia has been in the middle of the performance scale and India, of course, is the largest country in this area. Our initial conditions were not as favourable as in the two Asian cases. Although we introduced land reforms, they were only partially successful—certainly not as successful as in these two sets of countries. We have undertaken lot of infrastructure development—both physical and social—in the last 40 years. But we are nowhere comparable to the levels achieved in these countries. Saving rate in China, for instance, was as high as 35 per cent or even 40 per cent; our savings rate rose very much, no doubt, but nowhere comparable to this country. Our democratic framework has both strengths and limitations. Our problem is not that we are democratic, but that we have not deepened our democratic system sufficiently by taking it to the grass-roots level. Therefore, the governance system at the local level where the programmes have to be devised and implemented is not really effective. That is why the accountability of our governance, particularly in respect of rural development, is weak.

This has implications for designing economic reforms with a human face. The very first implication is that since our initial conditions were not as favourable as in the East Asian as well as South-East Asian situation, in designing our reform package we should take into account the need for building up infrastructure, physical as well as social, which was achieved there before introducing economic reforms. And since our socioeconomic structure remains inequitable, particularly in the rural areas, the distribution of benefits from development would be inequitable unless measures are taken to strengthen the asset base of the poor, improve their educational levels and ensure their participation in governance at the grass-roots level.

For achieving this, the United Progressive Alliance (UPA) government has come up with Common Minimum Programme, placing emphasis on employment guarantee, for example. The most important item highlighted in this programme is employment generation in our system. This is extremely relevant on which social scientists can contribute a great deal. According to the theory of liberalised factor markets, the cost of labour and cost of capital should be in conformity with the factor endowments. Since labour is abundant and capital is scarce, capital should be more costly, labour should become cheap so that employment generation could be higher. But why is that in our situation, despite liberalisation, employment generation has come down drastically? One may argue that this is because we have not introduced labour market reforms. The bulk of our labour force is in the rural areas where employment elasticity has come down drastically. This is an unorganised sector. The wages are already very low: even the statutory minimum wages are not implemented properly. So we cannot say that labour is costly. We do not have organised trade unions here. Even then why is the employment generation so slow? One should look into the whole structure of income distribution and the nature of demand for goods and services to see whether it generates sufficient demand for labour. One should also see how our subsidy structure is working. It is quite possible that excessive subsidies on irrigation and power are making our primary sector more capital-intensive. This is a challenging area of research for social scientists.

But economic reforms, to be successful, have to become participatory, especially at the grass-roots level, if we want to improve efficiency and improve natural resource management; it cannot be done without involving the people who are affected. So there is a lot of work in these two areas. Perhaps this is one of the most challenging areas of governance at the local level. We already have Water Users' Associations and Watershed Development Committees. This again is an area of research on institutions at the local level. There is a misapprehension that if more powers are devolved to local institutions, corruption may increase. This is an excuse for the reluctance to share power. On the other hand, the social activists who have been working in this area think that wherever power has been devolved, the management of resources has been more efficient and corruption has been checked because of local accountability.

Inaugural Address at the Conference on "Economic Reforms with Human Face" held at the Indian Council of Social Science Research (ICSSR), New Delhi, 13-14 August, 2004; Published in *Mainstream* XLIII(36), August 27, 2005.

3

Rural Transformation in China and India

Comparative Experience in the Post-Reform Period

IN China, economic reforms were initiated 27 years ago in 1978, while India began its reform process in the early 1990s. China recorded very high rates of overall economic growth as well as agricultural gross domestic product (GDP) growth and experienced a rapid reduction in poverty, in the post-reform period. China demonstrated that economic reforms need not slow down GDP growth, agricultural growth and reduction in poverty. On the other hand, India's GDP growth rate during its post-reform period, though slightly higher than in the pre-reform period of the 1980s, was less than expected, agricultural growth decelerated and poverty reduction was slower.

How has China been able to grow rapidly while India has not done so well? This is explained by the divergence in the initial or pre-reform conditions as well as in the political economy of economic reforms between the two countries. Let me first take up the initial conditions. In pre-reform China, the rate of capital formation was very high—over 30 per cent. As a result, China could develop much better physical infrastructure like irrigation, land development, roads, etc. In India, on the other hand, the rate of saving and capital formation was much lower. Secondly, the levels of social development like literacy level for males as well as females, and primary health care were much higher in China. Thirdly, unlike in India, landlessness was eradicated in China much before the introduction of economic reforms by implementing thoroughgoing land reforms.

However, China could not realise full benefits from the available physical and social infrastructure in the pre-reform period because of lack of incentives under the system of collectives and communes. When economic reforms were introduced in the late 1970s, terms of trade were turned in favour of agriculture and the communes were disbanded by introducing Household Responsibility System. These reforms released the latent energy of the people leading to the fuller utilisation of the potential in the system resulting in a very high growth and rapid reduction in poverty.

This brings me to the political economy of economic reforms in the two countries. China has been practising 'democratic centralism' as a political system. The rural social structure in China had undergone a radical transformation, as, unlike in India, the implementation of land reforms there was intertwined with the national liberation struggle. In India, on the other hand, the multi-class and multi-caste social structure of the colonial period remained intact at the time of independence. India adopted the parliamentary form of democracy without undergoing democratic revolution in the socioeconomic sphere.

Land reforms implemented in India after independence did result in the emergence of peasant proprietorship which was the single most important factor responsible for the success of the Green Revolution already in the 1970s and the 1980s. In this respect, India was perhaps a little ahead of China. India was also somewhat ahead of China in the 1970s in respect of the institutional infrastructure like rural credit and marketing. India's infrastructure of agricultural research and education is one of the best among the developing countries.

If, despite some of these favourable factors, India lagged behind China in respect of rural growth as well as poverty reduction, it can be explained basically by the other not-so-favourable initial conditions mentioned above and the political economy of economic reforms. Firstly, unlike the one party system under 'democratic centralism' in China, under India's multi-party system with parliamentary democracy, the political differences are aired freely and even the differences within the ruling party on issues concerning development are articulated openly and freely in the press. As a consequence, whereas quick decisions are taken in China, there are usually long debates and discussions before decisions are taken in India.

Secondly, the implementation of decisions taken is far more effective in China. This is because, even though market-oriented, it is a state-led rather than a 'market-led' economy. State power—political and bureaucratic—is backed by mass mobilisation in the implementation of decisions taken. In India, on the other hand, peoples' participation is tardy. Bureaucratic set-up in India, largely inherited from the colonial period, is highly rule- and control-oriented, and is enmeshed in lengthy procedures to ensure 'checks and balances' leading to tardy implementation of decisions taken. Besides, various interest groups in India, especially landed interests, are in a position to exert considerable pressures on the decision-making as well as implementation processes.

All this makes China a 'mobilising' society and India a 'debating' society, which explains why China is able to move much faster than India. But, should India, therefore, be resigned to the present slow growth syndrome? Or, alternatively, should India give up its parliamentary system of democracy in order to move faster? The answer is a clear 'No'. India cannot afford to remain a slow growing economy any longer. Further, it is neither possible nor desirable to give up its parliamentary system of democracy. The way out open to us is to work on our strength, *viz.*, political democracy, by turning it to our advantage, through deepening it and making it more participatory at the grass-roots level. Ironically, the unfinished 'democratic revolution' in India has to be completed through the democratic and peaceful means.

The amendments to our country's constitution, initiated by the late Prime Minister Rajiv Gandhi, to empower the local-level elected institutions by devolving to them more functions, resources and power over administration are already making some impact. Greater decentralisation from the centre to the states and from the states to the local governments can make the system more participatory, competitive and accountable to the people. Such a decentralisation combined with market incentives is the only feasible alternative to social mobilisation of China. Increased literacy and education and speeding up technical change, especially biotechnology and information technology, where the returns are very high, can become powerful engines of rural transformation resulting in high growth and poverty eradication. Already, state governments, local governments and non-governmental organisations (NGOs) in India are showing initiatives in this direction. Participatory watershed development, water users' associations and empowerment of women through their self-help groups (SHGs) are some encouraging examples.

India can learn a great deal from China in this respect. Any delegation of farmers and farm experts from India visiting China cannot but be impressed by the degree of decentralisation to the local-level institutions in the matter of decision-making as well as implementation, the flow of foreign direct investment (FDI) into rural enterprises and joint ventures with foreign enterprises. There is a need to promote such exchange of visits and collaboration between India and China for mutual advantage.

Keynote Address at the joint Conference of the Chinese Academy of Agricultural Sciences (CAAS) and the International Food Policy Research Institute (IFPRI) on "The Dragon and the Elephant: Agricultural and Rural Reforms in China and India" held in Beijing during November 10-11, 2003; published in Economic and *Political Weekly* XL(33), August 13-19, 2005.

4

Role of Sociopolitical Factors in Achieving Inclusive Growth

Comparative Experience of India and China

Introduction

Sociopolitical factors play a decisive role in determining the rate of growth of an economy, regional disparities in growth and how the benefits from growth are distributed between different social and income groups. This works through their impact on the mobilisation of resources and their allocation to different sectors and schemes as well as on the effectiveness with which the schemes formulated are implemented. India opted for parliamentary democracy with a multi-party system after independence but its multi-class and multi-caste social structure, inherited from the pre-independence period, remained largely intact except for the abolition of princely states and the intermediaries like the Jagirdars and the Zamindars. Its social structure has, however, been undergoing gradual change through the working of political democracy over the last 60 years. In China, on the other hand, the social structure underwent a radical transformation following the revolution in the late 1940s; the effective implementation of land reforms, which, unlike in India, was intertwined with its national liberation struggle; and nationalisation and public ownership of the major means of production. China adopted 'democratic centralism' as its political system which conferred pre-eminent status to the ruling Communist Party.

The avowed objective of economic planning, started in the early 1950s in both the countries, has been to step up growth in gross domestic product (GDP) and achieve equitable distribution of benefits from growth. It would be interesting to know the experience in achieving equitable or inclusive growth under these two contrasting sociopolitical systems during their plan periods spanning over half a century covering phases before and after they embraced globalisation by launching economic reforms. This paper is a modest attempt in this direction.

Growth, Poverty and Inequality

Even though China was better placed in agriculture than India, its per capita income in terms of parity dollars was substantially lower than in India in 1952 due to retarded manufacturing and service sectors. The rate of growth of National Income for the period 1952-1965 was about the same in the two countries: 3 per cent for China and 3.6 per cent for India. Even in 1986, that is, eight years after launching economic reforms, per capita income in China ($315) was only slightly higher than in India ($298)—five years before it was to launch economic reforms (Swamy, 1989). But by 2004, i.e., a quarter century after reforms in China and a little over a decade of reforms in India, per capita income in China rose sharply to $4660, which was almost twice the level of per capita income in India at $2460 (World Bank, 2008).

Both India and China have been able to reduce poverty as a result of steady growth in their per capita incomes. China's achievement in this respect has been far more impressive, much of the poverty reduction taking place in the early years of economic reform by the mid-1980s, when there was an exceptionally high growth in agriculture and equality in land holdings brought about by radical land reforms (Rao, 2005a). The remaining poverty in China is now concentrated mostly in the fast growing developed coastal regions owing to the high rate of in-migration from the poorer regions (Ghani, 2010), whereas in India, it is concentrated in the slow growing less developed central and eastern states. Apparently, the pull factors have been far more effective in inducing greater human mobility in China.

The pattern is the same concerning income inequality as in respect of growth: lower in China when compared to India in the pre-reform period but rising sharply and becoming significantly higher than in India well after the introduction of reforms. Inequality in the size distribution of income in China in the late 1970s was less than in India. It was in the urban sector that China exhibited impressive equality in incomes in the pre-reform period. In terms of poverty levels, illiteracy and health, China has done decisively better than India (Swamy, 1989). However, by 2004, income inequality in China, measured by the Gini Index (Gini Coefficient), rose to 0.47 as against 0.37 in India (World Bank, 2008). China has thus crossed the Gini Index of 0.40, a figure above which is usually considered 'dangerous' because of its potential for giving rise to social discontent and unrest (Weil, 2008).

It is interesting to see how India and China are placed on the world map in respect of income inequality. In the late 1990s, Denmark, Czech Republic, Japan, Norway, Slovak Republic and Sweden showed the lowest income inequality in the world with the Gini Index (Gini Coefficient) of around 0.25. The highest Gini Index ranging from 0.50 to 0.60 in the early years of the present decade related to the Latin American countries, e.g., Argentina, Bolivia, Brazil, Chile, Columbia and Peru. South Africa is also in the same group with the Gini Index of 0.58. The inequality in the United States was well below this level with the Gini Index of 0.41 (World Bank, 2008). Income inequality in India is thus well above that in the group of countries showing the lowest inequality and seems to be fast approaching the level in the United States—'the land of capitalism', whereas China—'the land of socialism'—already shows greater income inequality than the United States and is nearer the levels in some Latin American countries.

Sociopolitical Compulsions for Course Correction

In China, the dismantling of communes and absence of secure private ownership rights over land prevented farmers from benefiting from the rising economic value of land when the demand for land for non-agricultural purposes rose phenomenally under the impact of high overall economic growth as well as rapidly rising income inequality. With farmers becoming vulnerable to dispossession, there was a massive land grabbing, especially in fast-developing coastal areas, for the development of villas, golf courses and race tracks

(Riskin, 2008). As many as 40 million farmers lost their land within a decade and the incidence of slave labour, including children, surfaced in a big way (Chandra, 2009). Besides, throughout the 1990s, most of the state and collectively owned enterprises were privatised which led to the millions of workers being laid off. Further, dismantling of rural collective economy and basic public services forced millions of peasants to migrate into cities as workers (Li, 2008). Despite such massive migration into cities, regional as well as rural-urban disparities in incomes became greater. Widening social polarisation led to strikes, protests, confrontation over corruption, land takeovers and environmental damage (Lotta, 2009). The number of such collective disturbances rose 10 times from 8700 in 1993 to 87,000 in 2005 (Lee and Selden, 2008).

In India, the rate of growth of agriculture GDP which was a little over 3 per cent per annum when the Green Revolution was in full swing in the 1980s decelerated during the Ninth and Tenth Plan periods, with the growth in food grains output falling below the population growth rate. The difference in per capita income between agriculture and non-agriculture widened sharply. The regional divide became similarly sharp with the per capita gross state domestic product (GSDP) of Bihar—the poorest state in the country—dropping to 20 per cent of that in the richest state in 2004-05 (Planning Commission, 2008).

Since the less developed states in India account for over 60 per cent of the population below the poverty line in the country, the pace of poverty reduction slowed down in the post-reform period, despite the rise in average national per capita GDP at a faster rate. The popular discontent arising from this economic polarisation reached its climax seven years ago when the GDP growth rate was at its peak and 'India Shining' campaign, then in full swing, coexisted with widespread rural distress and suicides by a large number of farmers (Rao, 2009).

Despite lower per capita income, higher poverty and democratic institutions providing outlets for ventilating grievances, the social unrest manifest in India appeared less serious than in China at the time of the formulation of their respective Eleventh Five Year Plans.

This could well be attributed to the fact that China was ahead of India by at least a decade in implementing market-oriented reforms and experienced a much higher growth rate in GDP than India for a quarter century with sharply rising income inequalities accompanied by large-scale dispossession and displacement of vulnerable groups. This experience shows that high inequalities are no less a source of social discontent than high incidence of poverty. Even so, the mass discontent in India did appear to cross the 'tolerance limits' within a parliamentary democracy, as is evident from the change of government following the general elections in 2004.

Rising income inequalities, dispossession, loss of traditional livelihoods, inadequate measures for compensation, rehabilitation and resettlement, and emerging social tensions forced both India and China to attempt a course correction in their policies, surprisingly around the same period, in the formulation of their respective Eleventh Five Year Plans. India's Eleventh Plan (2007-2012) promised 'Faster and Inclusive Growth' through 'reforms with human face', and China's Eleventh Plan (2006-2010) came out with the objective of 'Building Harmonious Society' by 'putting people first'. Unlike in India where the multi-party system with a parliamentary democracy can force a change in the policy framework relatively quickly in response to the emerging unrest, the delayed response to the discontent in China could well be attributed to its political system which assigns a pre-eminent place to a single party.

Structural Bottlenecks to Inclusive Growth

Notwithstanding the recent attempts at course correction in both the countries towards 'inclusive growth' and 'harmonious development' through their Eleventh Five Year Plans, it would be naïve to expect that the change would be quick and smooth. The recent official pronouncements in both the countries represent, at best, the avowed intentions of the ruling elite made in response to the prevailing discontent. The roadblocks to achieving inclusive growth are formidable as they are structural in nature—having to do with political system in China and social structure in India.

Political System in China

China embarked upon economic planning in the early 1950s with favourable initial conditions in respect of its social structure that emerged as an aftermath of the revolution. When it launched economic reforms in the late 1970s, it had the additional advantage of adequate physical and social infrastructure built over a period of three decades of economic planning. Its authoritarian political system and mobilising power deriving from the highly organised and centralised party apparatus right from the grass roots enabled the authorities to take quick decisions and implement the development schemes effectively (Rao, 2005b). This is amply borne out by the unprecedented high rate of growth in GDP achieved by China over the three decades of reform. But the equally unprecedented rise in income inequality in this period is something not quite anticipated by Marxist ideologues. At the time of launching economic reforms, 'egalitarianism' practised earlier was openly derided officially and a call was given to the people to become rich: "Let some people get rich, so that others can follow" (Rao, 2005c). This is surprising because with all the variants of 'revisionism', one never found such a slogan in communist literature. Evidently, they became sick of the extreme form of 'egalitarianism' practised earlier and the resultant slow growth in productivity. They admit that economic disparities would rise under the new policy, but believe that these can be corrected in course of time.

If income inequalities in China rose sharply to the Latin American levels within a short period of 2-3 decades despite an equitable social structure prevailing at the onset of planning and the development of relatively broad-based physical and social infrastructure before launching economic reforms, a plausible explanation is the prevalence of a political system which confers pre-eminent status to the ruling party with hardly any statutory and organised political opposition. This could be the source for the emergence of high inequalities in the wake of globalisation and market-oriented reforms.

The emergence of inequalities is often a consequence of high growth owing to the differential comparative advantages experienced by different regions arising from geographical and other locational factors, and to the inter-personal differences in access to education, skills and resources. An analysis of income inequalities in China concludes that the "most important factors explaining overall inequality are differential returns to schooling and sector of employment ... investing in people, in particular improving education service in lagging areas to poor people are important for sustainable growth and equitable distribution in the long-run" (Luo and Zhu, 2008). Neglect of the development of education and skills in the lagging regions and sectors and among the poor is traceable to the biases in decision-making regarding the allocation of resources and other measures. The sensitivity and resilience of the prevailing political system to take timely corrective measures could make a big difference to the degree of inequalities and mass deprivation.

The observations of Milovan Djilas, made decades before a country like China under 'communism' embraced globalisation and market reforms, are prophetic, especially in respect of the consequences of an authoritarian and closed polity for income distribution:

> The new class obtains its powers, privileges, ideology, and its customs from one form of ownership—collective ownership—which the class administers and distributes in the name of the nation and society ... the so-called socialist ownership is a disguise for the real ownership by the political bureaucracy (Djilas, 1957).

The collective or socialist ownership is no longer a predominant feature of Chinese economy. In fact, the state economic sector accounts for only 35 per cent of economy, which, nevertheless, serves as an economic base of power of the leading Chinese bourgeoisie (Lotta, 2009). It is a 'state-led' market economy where, contrary to conventional wisdom, state plays a pre-eminent role in influencing the growth rate as well as the distribution of income. In the post-liberalisation period, the patronage extended by the state in awarding contracts, choice of location for private sector projects and technical institutions, decisions about the number, type and location of special economic zones, land acquisition and compensation policies, and various kinds of patronage extended to different enterprises and activities, could together make a greater impact, when compared to the pre-liberalisation period, on the growth of economy as well as on how the benefits from growth are distributed. If collective ownership of means of production was the source of power and privileges for the ruling political leadership and bureaucracy in the socialist countries half a century ago, it is basically the continued exclusiveness in access to state power conferred by the prevailing political system that enables the ruling political and bureaucratic class to distribute patronage to the privileged groups of people for operating in a market economy. This could well have become a source of high inequalities—inter-regional as well as inter-personal—in endowments of infrastructure, skills and resources.

The Chinese leadership has seen growing "inequality as a central political threat to political stability, that is, to communist party rule ... and in 2005 re-emphasised the primacy of social justice ... to construct 'harmonious society'" (Lee and Selden, 2008). This is evident from the use of redistributive instruments like fiscal transfers to reduce regional disparities. Although regional disparities continued to increase, they levelled off after 2000 (Wang, 2008). However, the change required is really basic, *viz.*, reforms in the political system that "engages broad democratic forces in the business of policy making" (Riskin, 2008). But, "the general yearning for greater democracy and transparency is almost entirely blocked" and there is "growing frustration at the inflexibility of the current system"(Weil, 2008).

The prevailing authoritarian and closed political system appears to be a stumbling block to inclusive and ecologically sustainable development. Despite periodic attempts at course correction, prospects for inclusive development look gloomy without systemic political reforms towards an open and competitive polity.

Social Structure in India

Unlike China, India started its economic planning within a highly inequitable social structure in the early 1950s. The level of infrastructure—physical as well as social—has been way below that in China, as the saving and investment rates have been much lower than in China. Even now, social structure in rural India is more inequitable than in China. The Gini Index of land distribution in rural India was 0.74 in 2003, as against 0.49 in China in 2002—which, however, was zero after the revolution. The Gini Index of the distribution of adult schooling years in the population in the year 2000 in India (0.56) was higher than in China (0.37). It was higher than even in most Latin American countries (World Bank, 2006). The pace of poverty reduction in India is slower than in China, "not just because growth has been much faster in China, but also because the same 1 per cent growth rate reduces poverty in India much less, largely on account of inequalities in wealth—particularly land and education" (Bardhan, 2007).

Like China, Indian economy is also a 'state-led' market economy and, contrary to common belief, the role of the state has become critical in determining the course of development even in the post-reform period. As in China, its direct role in the allocation of investment resources has declined, but its indirect role as facilitator of private investment and economic activity has vastly increased, with greater impact on growth and distribution of benefits than in the pre-liberalisation period. If the change is not as rapid as in post-reform China, it is because of various in-built pressures, checks and balances in the decision-making process, and the hierarchical social structure which has a debilitating influence on the delivery systems, resulting in much lower outcomes than expected.

Although India's traditional inequitable social structure has been undergoing gradual transformation through the working of political democracy, the impact of entrenched economic interests represented by the big landed sections, business groups and growing middle class on economic policy making as well as on changes in government has been significant. Being a mixed economy, private sector has had a strong presence in India since the beginning of economic planning. Its role has considerably increased with the opening up of the economy following economic reforms. These entrenched interests have a dominating influence in policy making even when the governments come to power with popular mandate to address the concerns of the poor like employment guarantee, food security and other measures for social protection.

The recent experience demonstrates how the Eleventh Plan's agenda of inclusive growth is being undermined in practice by the pulls and pressures from such interests. For example, an avowed objective of the Eleventh Plan is to reduce regional disparities in development. But, whereas the fiscal transfers recommended by the statutory body like the Finance Commission are significantly higher in per capita terms for the poorer states, the discretionary schemes of the central government show higher per capita expenditures in the richer states. Similarly, subsidy on fertiliser benefits richer regions much more than the poorer regions (Ghani, 2010).

At the moment, the disadvantaged sections are basically made to serve as 'vote banks' through the dispensation of patronage by the privileged groups who, being effective decision makers, are the dominant players. Mere electoral participation by the underprivileged cannot be equated with participatory democracy. In the ultimate analysis, empowerment of the disadvantaged sections means their ability to actively participate in decision-making and implementation by becoming the dominant players.

This cannot be left to the forces that are random and chaotic. There needs to be a clear agenda for speedy social change. The basic goal has to be the political and economic empowerment of the disadvantaged sections whether by caste, religion, gender or regions.

Conclusion

In China, there are indications that "activism informed by notions of rights and citizenship, as well as class inequality, has the potential to heighten pressure for change in the social order" (Lee and Selden, 2008). How soon the change will occur in the political system in China and how smooth the transition would be is uncertain. But the change seems inevitable because individual freedom in economic decision making within a market setting is incompatible, in the long-run, with the denial of individual freedom in the political sphere (Chakravarty, 1993).

India's open and democratic polity provides opportunities for public dissent, protests and periodic change of governments. Of late, a large number of 'development projects' in different parts of the country which displace the poor and destroy their livelihoods and harm the ecology have been stalled because of the protests and organised resistance by the affected people. Therefore, despite its shortcomings,

India's democratic framework constitutes the greatest asset as an instrument for achieving inclusive and sustainable development.

The social change underway in India is faster than before—thanks to the rise in awareness of the common people, their improved staying power and growing civil society activism. But it is bound to be gradual because of resistance from the vested interests. It is also bound to be, by and large, peaceful because of the visible prospects for change through the democratic process and pro-poor measures, even in areas presently affected by extremist violence. There is already an air of optimism about such a change.

References

Bardhan, Pranab (2007). "Inequality in India and China: Is Globalisation to Blame?", *Yaleglobal*, 15 October.

Chakravarty, Sukhamoy (1993). *Selected Economic Writings*. Delhi: Oxord University Press.

Chandra, Nirmal Kumar (2009). "China and India: Convergence in Economic Growth and Social Tensions?", *Economic & Political Weekly*, January 24-30.

Djilas, Milovan (1957). *The New Class*. London: Thames and Hudson.

Ghani, Ejaz (2010). *The Poor Half Billion in South Asia*. Oxford University Press.

Lee, Ching Kwan and Mark Selden (2008). "Inequality and Its Enemies in Revolutionary and Reform China", *Economic & Political Weekly*, December 27.

Li, Minqi (2008). "Socialism, Capitalism, and Class Struggle: The Political Economy of Modern China", *Economic & Political Weekly*, December 27.

Lotta, Raymond (2009). "China's Rise in the World Economy", *Economic & Political Weekly*, February 21-27.

Luo, Xubei and Nong Zhu (2008). *Rising Income Inequality in China: A Race to the Top*. The World Bank, East Asia and Pacific Region, Poverty Reduction and Economic Management Department. August.

Planning Commission (2008). *Eleventh Five Year Plan (2007-2012)*, Vols I & III, New Delhi.

Rao, C.H.H. (2005a). *Agriculture, Food Security, Poverty, and Environment: Essays in Post-Reform India*. New Delhi: Oxford University Press.

———. (2005b). "Rural Transformation in China and India: Comparative Experience in the Post-Reform Period", *Economic & Political Weekly*, August 13-19.

———. (2005c). *Agricultural Growth, Farm Size and Rural Poverty Alleviation in India*. New Delhi: Academic Foundation.

———. (2009). "Inclusive Growth: Recent Experience and Challenges Ahead", *Economic & Political Weekly*, March 28- April 3.

Riskin, Carl (2008). "Property Rights and Social Costs of Transition and Development in China", *Economic & Political Weekly*, December 27.

Swamy, S. (1989). *Economic Growth in China and India 1952-1970*. New Delhi: Vikas.

Wang, Shaoguang (2008). "Double Movement in China", *Economic & Political Weekly*, December 27.

Weil, Robert (2008). "A House Divided: China after 30 years of 'Reforms'", *Economic & Political Weekly*, December 27.

World Bank (2006). *World Development Report 2006: Equity and Development*. Oxford University Press.

———. (2008). *World Development Indicators*. Washington, D.C: World Bank.

This paper is based on the Silver Jubilee Lecture at the 27th Annual Conference of the Andhra Pradesh Economic Association held at the National Institute of Technology (NIT), Warangal, 14-15 March, 2009; Published in *Economic and Political Weekly* XLVI (16), April 16-22, 2011.

5

Globalisation and Inequality

Introduction

Globalisation has been progressing at a fast pace over the last two to three decades. This is manifested, in the first place, by the fast growth in global trade—a five-fold rise in real terms since 1980. Secondly, financial globalisation has been rapid. For example, cross-border financial assets as a ratio to world gross domestic product (GDP) more than doubled during this period. Third, technological change associated with trade and foreign direct investment (FDI) has also been rapid (IMF, 2007).

Globalisation represents external liberalisation, but over the same period major changes in development paradigms led to internal liberalisation of economies. There has been a decline in the role of the state and centralised planning with an increase in the role of the market and the private sector which had a significant impact on inequalities. Thus, the observed changes in inequality over this period are the combined result of external as well as internal liberalisation or the so-called structural adjustment.

These changes in development paradigm in India are reflected in a decline in public investment in the economy, including in agriculture, from around 10 per cent of GDP in 1980 to around 7 per cent in 2005-06. The growth of public expenditure on social sectors and access to institutional credit slowed down leading to a significant rise in income inequalities.

Income Inequalities

Inter-country inequalities in per capita income have shown a slight decline in the recent period. But, if China and India are excluded, inter-country income inequalities are rising, as the low income countries in Sub-Saharan Africa (SSA) are showing a decline in incomes. However, if income inequalities within these countries, especially in large countries like China and India, continue to increase, the world inequality may also increase because of the large weight of China and India in the world economy (World Bank, 2006).

Globalisation has been marked by the rise in GDP growth rates except in some African countries. Income inequalities in the United States (US) have been rising since 1970s, and dramatically so in the United Kingdom (UK) since 1980s. In East Asia, inequality rose faster in the second round of globalisation when compared to the first round characterised by the 'Asian miracle' of 1960s and 1970s. In China, income inequality is greater than in India and has been rising faster. Rising Gini coefficients in this period are explained largely by the increasing share of richer quintiles, signifying the rapid rise in the number of billioneers (World Bank, 2006).

Nexus between Growth, Inequality Increase and Poverty Reduction

Experience of a large number of countries e.g. India, China, Viet Nam, Bangladesh, Ghana, South Africa, Mexico, Chile, etc., shows that, unlike the east Asian countries in 1970s and 1980s, no country achieved growth without increasing inequality: growth, yes, but higher at the top than at bottom. Poverty has been declining, but at a lower pace than before, i.e., growth elasticity of poverty reduction has declined because of the rise in income inequality (Kanbur, 2008). In India, poverty (below ₹ 12 per capita consumption per day in 2004-05) came down significantly, but the poor and the vulnerable defined as those showing below ₹ 20 per capita consumption per day in 2004-05 constituted as high as 77 per cent of total population (NCEUS, 2007).

Sources of Inequality (From Globalisation)

Trade liberalisation, by and large, has equalising effect in developing countries because of the rise in labour-intensive exports, especially from agriculture, leading to a rise in wages of low-skilled labour, which is abundant in these countries. The situation could, however, be opposite in developed countries where high-skilled labour is abundant: wages of skilled labour may decline causing a rise in income inequality (IMF, 2007).

Rise in FDI in developing economies may raise growth as well as inequalities. Because access to finance is biased towards higher income and asset groups, especially in developing countries, tapping global savings may accentuate inequalities.

Information and communication technology (ICT) is associated with trade and FDI (and financial liberalisation) and is thus not independent of globalisation. ICT rose sharply in the process of globalisation and is the main driving force behind rising inequalities. This is because such technical changes typically raise the demand for skilled labour which is in short supply in these economies and is highly unequally distributed between different regions and social groups (IMF, 2007).

Internal Liberalisation and Inequality in India

With the decline in the role of planning *vis-à-vis* markets in India in the wake of globalisation, there has been a decline in public investment in agriculture, rural infrastructure and in the less developed regions. This has led to the accentuation of inter-state as well as rural-urban disparities in incomes. Further, owing to a decline in social sector expenditures, especially in the immediate post-liberalisation period, inequalities in the access to education have increased. There has been a reduction in the number of bank branches in rural areas in this period. Reduced flow of institutional credit to rural areas has adversely affected weaker sections such as small and marginal farmers and small and medium enterprises.

Relevance of Sociopolitical Factors

Inequalities in wealth and income lead to unequal access to power, status and influence. Thus, globalisation and internal liberalisation of the economy within an inequitable social structure result in accentuating the existing inequalities in income and wealth in the society. Following liberalisation of the economy, the direct role of the state as investor may decline but its indirect role continues to be important and may even increase towards patronising the private sector. This is because the character of the State itself may undergo changes with increasing influence being exerted by the economically powerful sections on the levers of state power. This is especially so under dictatorships or the totalitarian political systems. This may indeed explain why income inequalities seem to have increased faster in China than in India in the post-liberalisation period. Democracies may be less prone to such state capture than dictatorships but even democracies have proved to be vulnerable to pressures from the powerful private interests, particularly when the prevailing social structure is inequitable as in India.

Remedies

Narrow and non-inclusive financial sectors are inequitable and inefficient. Advantages of globalisation notwithstanding, global markets are far from equitable. Less voice in the governance of the global institutions and unequal endowments limit the ability of the developing countries to benefit from globalisation. There is, therefore, a need for greater participation of the developing countries in international rule-setting. Towards this end, reform of multilateral institutions has to be high on the agenda. It is also imperative for the developing countries to build up their own endowments or capabilities.

Policy action on the domestic front turns out to be central to reducing inequalities in income and wealth in the wake of globalisation and internal changes in development paradigm, especially because of the overriding importance of the initial conditions such as domestic socioeconomic structure in influencing the distribution of gains from globalisation and liberalisation. The major areas needing policy action are: development of physical infrastructure that is regionally broad-based; expanding the opportunities for education and for acquisition of necessary skills in the less developed regions and for socio-economically disadvantaged sections; and ensuring financial inclusion through innovative methods of credit delivery.

References

International Monetary Fund (2007). *World Economic Outlook-Globalisation and Inequality.*

Kanbur, Ravi (2008). *Poverty and Distribution: Twenty Years Ago and Now.* Fifth Lecture in Golden Jubilee Series. Institute of Economic Growth, University of Delhi.

National Commission for Enterprises in the Unorganised Sector (NCEUS) (2007). *Report on Conditions of Work and Promotion of Livelihoods in the Unorganised Sector.* New Delhi.

World Bank (2006). *World Development Report 2006: Equity and Development.* Oxford University Press.

Presentation at the National Seminar on 'Globalisation and Inequality' organised by Bengal Economic Association in collaboration with the Indian Economic Association & CSESD, RBU, at Rabindra Bharati University, Kolkata, 20 June, 2009. See Debes Mukhopadhayay (ed.) (2011). *Globalization and Inequality*, New Delhi: Deep & Deep Publications Pvt.Ltd.

6

Approach to the Eleventh Five Year Plan

A Comment

THE Planning Commission has to be guided by the politically mandated objectives but, as an expert body, it can present feasible alternatives for achieving these objectives and spell out the tasks ordained by such alternatives. One has to bear in mind that the United Progressive Alliance (UPA) government came to power with a popular mandate which can rightly be summed up as ' More Inclusive Growth'.

Against this backdrop, the Planning Commission has done a commendable job by pointing out that a gross domestic product (GDP) growth rate of around 8.5 per cent is feasible considering the recent performance of the economy as well as its potential. Equally persuasive is its argument that such a growth is necessary to generate the resources for investment as well as the demand for achieving 'inclusive growth'. For example, agricultural growth of 4 per cent and the high growth of rural non-farm enterprises require massive public investments in rural infrastructure, and this high rural growth can be sustained only if adequate demand is generated in the economy through higher GDP growth.

Also noteworthy is the Commission's candid expression of its reservations on the Fiscal Responsibility and Budget Management (FRBM) Act which requires bringing down revenue deficit—which currently covers several asset-creating expenditures—to zero by the year 2009. Since this could unduly restrict plan outlays with large 'revenue' component, the Commission makes out a case for amending our FRBM legislation to conform to international practice, where the focus is on controlling the fiscal and primary deficits and not on revenue deficit.

The Planning Commission could have gone further to emphasise that, in the past, slippages in realising targets in respect of sectors bearing on inclusive growth, e.g., agriculture, poverty reduction, primary health, elementary education and reducing regional disparities in development, etc., have been far more serious than the slippages in respect of GDP growth. For example, in the post-reform period when there was an acceleration in GDP growth, there was a deceleration in the growth of agriculture and the pace of poverty reduction has been slower.

Considering the prevailing socioeconomic structure and the nature of our polity, the challenges for achieving more inclusive growth are much greater than for stepping up the GDP growth rate. The Planning Commission could have drawn attention to such difficult tasks.

Of the various 'divides' enumerated by the Commission, the rural-urban and regional divides seem, in a way, to be central as they considerably overlap with other divides such as rich-poor and employed-unemployed, so that bridging the first two would significantly dent on the other divides. The gender divide, however, is all pervasive and requires to be addressed comprehensively in every sphere.

For bridging the rural-urban divide, the Commission envisages more than doubling of agriculture growth from below 2 per cent to 4 per cent and stepping up the growth of rural non-farm enterprises. The major avenues of agricultural growth known are three: irrigation, technology and other support systems like credit, extension, marketing and instruments for input delivery. The last category of measures offers relatively quick results from the efforts made and the UPA government can take credit for some tangible results on this front.

It is proposed to bring, through Bharat Nirman programme, 2.5 million ha per annum under irrigation as against about 1.5 million ha in the recent past. Some indication of the regional profile of this proposed addition would be useful because the gap between the irrigation potential and its exploitation is highest for states in central and eastern India which lack the necessary resources to match central assistance under Accelerated Irrigation Benefit Programme (AIBP).

The strategy for rainwater harvesting and ground water recharge in rainfed areas envisages setting up the National Rainfed Areas Authority. This superstructure can yield the desired results only if the focus on participatory approach at the grassroots level is revived and the drive for restoration of tanks is integrated with the ongoing watershed development programme.

The Commission admits that there is no technological breakthrough in sight in the next five years. As such, it seems to rely rather heavily on making good the 'knowledge deficit'—the gap between the proven technology and its use in the field. It is not clear which crops and regions are likely to benefit from bridging this gap. As it is, crop yields in north-western India are often found to be higher than those in the experimental stations.

Although the planners have always targeted for 4 per cent agricultural growth, the highest rate achieved over a longer period of a decade was only a little over 3 per cent during the 1980s when the annual addition to irrigation was 2.5 million ha; high-yielding technology for wheat as well as rice spread fast in eastern India, and the institutional support, e.g., credit and extension, was also at its best. Whereas agricultural growth of 4 per cent is necessary, one would look for a more credible provision of public investment and institutional support for achieving this growth in areas where the potential exists.

Apart from textiles, food processing offers a great promise in the rural non-farm sector. Experience in East Asia also bears this out. In this respect too, credible initiatives are needed for attracting private investment through infrastructure development and for encouraging contract farming to cope with the risks inherent in the diversification of agriculture.

The Mid-Term Appraisal of the Tenth Plan had contained an incisive and bold analysis of growing regional divide. The minimum the planners could aim at is narrowing down the divergence in inter-state growth rates in gross state domestic product (GSDP), if not bringing down the disparities in per capita income. But the Approach document is silent on the goal.

Following the recommendations of the Twelfth Finance Commission, the loan component of Central Assistance for state plans under the Gadgil Formula is done away with and the states are left to the market for accessing resources for plan investments. It is not clear whether the Centre would mediate in the market for accessing the resources required for the backward states or extend assistance through other mechanisms.

The only window mentioned is the Backward Area Grants Fund specifically designed for the development of selected backward districts. But plan spending at the district level even in the best of situations does not exceed 30 per cent of total state plan outlay, the remaining being allocated at the state level. The expenditure incurred under the Bharat Nirman Programme so far does not show any tilt towards backward states. A credible strategy is needed for building sufficient infrastructure in the backward states and in the backward regions within some of the larger states.

Economic Times, August 31, 2006.

7

National Rural Employment Guarantee

Tasks Ahead

THE sharp deceleration in the growth of employment, particularly in rural areas, during the 1990s gave rise to popular discontent which was overwhelmingly reflected in the verdict of the electorate in the general elections of 2004.

If wailing over the 'wasteful' use of resources on the Employment Guarantee Scheme (EGS) is pointless—as it has little justification in the present context of rural distress—any attempt to underestimate the challenges facing the effective implementation of the Scheme is misplaced. Except as a relief operation in conditions of drought or famine, provision of employment through public works has never been regarded as the first best option. For a developing economy like ours, employment-oriented growth has always been regarded as the first best alternative for generating productive employment. Now that the EGS is a reality, it is time for some serious introspection on the part of the planners and politicians with a view to reversing the on-going jobless growth as well as for exploring bold and innovative methods for the effective implementation of the EGS.

If our agricultural output did in fact grow at a minimum rate of 4 per cent per annum, as targeted by the successive Five Year Plans, together with large-scale diversification of agriculture into dairying, horticulture, fisheries, etc., and if there was a significant growth in the rural non-farm sector, comprising agro-processing and other rural industries, the employment growth in rural areas would have been sufficient to absorb bulk of the growing labour force and there would not have been any need for the EGS.

Maharashtra provides a classic illustration of how the introduction of EGS became inevitable in the 1970s because the growth rate of agriculture there fell persistently short of population growth over a long period. Despite the notable achievements of EGS, the persistence of rural poverty in Maharashtra, even after the implementation of EGS for three decades, underscores the importance of employment-oriented growth for sustainable reduction in poverty.

There has been a sharp deceleration in the growth of crop output in the country in the post-reform period of 1990s and a very slow pace of diversification of agriculture and growth of rural non-farm enterprises. This retrogression is attributable to the continued decline in public investment and effort for building up rural infrastructure, e.g. irrigation, rural roads, agricultural research and extension and institutional credit. Therefore, what is at issue is not the need for economic reforms *per se* but their proper sequencing by according high priority to agriculture and rural industries in the initial stages of reform when there is widespread incidence of unemployment and poverty.

As rightly enunciated in the National Common Minimum Programme (NCMP), agriculture and rural infrastructure development should top the national agenda. At the same time, effective implementation of EGS should be an area of major concern. The objective in this regard should be two-fold: to use the Scheme for productive activities which have the potential for sustained employment generation and to minimise leakages of benefits from the Scheme to the non-poor.

The best way to ensure high productivity of public works in rural areas is to coordinate and integrate EGS with the on-going programme for 'Bharat Nirman' involving the development of rural infrastructure like irrigation, roads, water supply, rural electrification, housing, etc. As it is, under the Food-for-Work programme, currently in operation in 150 most backward districts of the country until the EGS comes into force, eco-development works such as soil and moisture conservation are to be given a high priority.

Integrating EGS with Bharat Nirman, while retaining its identity as well as dedicated funding for the Scheme, would pose a major challenge to the planning process right from the Planning Commission to the grass-roots level. This would require detailed planning of proposed works at the local level through coordinated effort between concerned departments including the one implementing the EGS. The estimated requirement of labour for all such works in different seasons against the availability of labour for work in each area would give an idea of the possible absorption of labour in works under Bharat Nirman. Arrangements can be worked out to channel labour into such works on a preferential basis through appropriate incentives where necessary.

However, there would be areas where works under Bharat Nirman are either not being undertaken or are located too far away for local labour to move into. Those in charge of EGS should be prepared with their plans for productive works to meet the demand for work upto 100 days per household as envisaged in the Scheme. If the coordination between different departments is effective, the actual demand for work from the EGS can be expected to be much lower than when there is no such coordination.

Certain features of the Scheme, *viz.*, the responsibility for implementation entrusted to *panchayats*, invoking of social auditing through *gram sabhas* and the possibility of using the newly enacted Right-to-Information Act, hold promise for minimising the leakages from the Scheme. However, since the risk of leakages arises essentially from the top-down nature of the Scheme (notwithstanding the involvement of *panchayats*) and the prevailing hierarchical social structure that is exploitative, unless the political parties rise above partisan considerations and display the same unity of purpose as at the time of passing the Bill, it would be difficult to prevent the present Scheme from quickly degenerating into the likes of the old.

After all, we have had a plethora of rural employment schemes over the past three decades, some of them even entrusted to *panchayats*, the experience from which has only served to breed cynicism. While all-out efforts need to be made to reverse this trend, it would be naïve to envisage EGS as the main instrument for poverty eradication throughout the country. EGS quickly graduating into a programme for infrastructure development as an integral part of Bharat Nirman provides an efficient and the least risky way for sustainable rural development and poverty eradication.

Economic Times, September 5, 2005.

8

Growing Regional Disparities in Development in India

Post-Reform Experience and Challenges Ahead

PROF KHUSRO was a versatile economist. His academic interests ranged from subjects like agriculture and rural development to monetary economics and macroeconomic policy. His writings in the 1970s and the 1980s were highly critical of the economic policies in that period, and anticipated economic reforms introduced in the country in the early 1990s. But, all through, he remained a votary of economic growth with social justice. As a member of the Planning Commission in the early 1980s, Prof Khusro was in-charge of the Division on Financial Resources. Later, as the Chairman of the Eleventh Finance Commission, he grappled with issues concerning Centre-State Financial Relations in India. It is interesting to note in this context that the Award of this Commission was distinctly favourable to the less-developed states. In fact, the inverse (rank) correlation between per capita income of states and per capita transfers was highest (-0.97) in the case of Eleventh Finance Commission Award (Singh, 2005).

Growing Regional Disparities in Development

It is now well established that the inter-state disparities in the growth of gross state domestic product (GSDP) have increased in the post-economic reform period beginning from the early 1990s when compared to the 1980s. In general, the richer states have grown faster than the poorer states (Ahluwalia, 2000; Dev and Ravi, 2003; Bhattacharya and Sakthivel, 2004). The regional disparities in per capita GSDP growth are even greater because the poorer states in general have experienced a faster growth in population. The exercise on Regional Equality presented by the Planning Commission in the 12th Plan confirms this position, despite the recent acceleration in the GSDP growth rates of some poorer states, including especially Bihar, and rise in the ratio of their average growth rates to the all-India average growth rate (Planning Commission, 2012).

Although these disparities have accentuated in the post-reform period, they have been building up in the pre-reform period itself. For example, according to Central Statistical Organisation (CSO) data, in the early 1960s, the per capita GSDP of the richer states like Punjab, Maharashtra and Gujarat was, on an average, about 80 per cent higher than the average per capita GSDP of the bottom four states *viz.*, Bihar, Uttar Pradesh, Orissa and Madhya Pradesh. This disparity increased to 125 per cent by the early 1970s (Rao, 2005) and was contained at a little over 100 per cent during the eighties, but escalated steeply to 156 per cent towards the mid-1990s (1993-96), and further to 176 per cent during 2007-10.

States whose per capita GSDP is below the national average together account for over 60 per cent of the country's population and as high as 75 per cent of the country's population below the poverty line. Further, these states account for nearly 60 per cent of the population belonging to the socioeconomically disadvantaged sections like scheduled castes (SCs) and scheduled tribes (STs). There is thus a large potential for growth which needs to be exploited for sustaining development in the country over a long period. This is necessary for improving regional and social equity and for strengthening national integration.

Sectoral Composition of Growth

How does one explain the accentuation in inter-state disparities in development in the post-reform period in the country and, indeed, the tendency of a gradual build-up in these disparities over the plan period? One way to study is to decompose by looking at the emerging regional disparities with respect to the individual sectors, i.e., the Primary, the Secondary and the Tertiary. The inter-state variation in per capita GSDP originating from the Primary Sector, measured by the coefficient of variation, has always been lower than that from the Secondary Sector GSDP. There is also no evidence of an increase in its variability across regions in the post-reform period when compared to the 1980s. This is understandable because agriculture is based essentially on land and labour which are widely distributed across the country. There are, no doubt, significant regional disparities in the availability of physical and institutional infrastructure for agriculture like irrigation, rural electricity and institutional credit. However, despite these constraints, technological changes in Indian agriculture represented by the Green Revolution were adopted widely in the country, including especially the eastern region, by the end of the 1980s.

The GSDP per capita from the Secondary Sector, on the other hand, shows much higher variability across regions and this variability rose significantly in the post-reform period. The Tertiary Sector which has been the prime mover of GDP growth in the post-reform period has generally grown faster in the industrially advanced states and shows a significant rise in its variability across states during this period. We know that there has been a steep reduction in the share of the Primary Sector in the overall GDP, a modest rise in that of the Secondary Sector and a big rise in the share of the Tertiary Sector. Thus, the decline in the relative importance of the Primary Sector—which shows lower disparities—and the rise in the importance of the other two sectors showing higher variability explain, at the compositional level, the rise in inter-state disparities in per capita GSDP in the post-reform period.

Whereas inter-state disparities in per capita GSDP have increased significantly in the post-reform period, the corresponding disparity in Human Development Index (HDI) has declined. The coefficient of variation in HDI declined from 0.22 in 1981 to 0.19 in 1991 and 0.16 in 2001. This could be because investment requirements in primary education and primary health care are not as heavy as in the capital-intensive physical infrastructure, and the results in terms of improving literacy rate and the life expectancy at birth are relatively quick. However, the rise in public expenditure in these sectors has been very slow, especially in the poorer states in the post-reform decade (Dev and Mooij, 2003). This is reflected in the slower decline in regional disparities in HDI in the post-reform decade. Since human development has a positive impact on GDP growth, inadequate attention to this sector in the poorer states in the immediate post-reform period could also be responsible for the rise in regional disparities in per capita GSDP.

The estimates from the India Human Development Report 2011 show that the coefficient of variation of the HDI for states has fallen sharply between 2000 and 2008, the improvement in HDI for majority of low per capita income states being greater than for India as a whole (Planning Commission, 2012). This convergence in HDI, by itself, can be expected to have a softening impact on inter-state disparities in per capita GSDP in course of time.

Causal Factors

Investment

Planning and Public Investment

Investment in physical and human capital, technical change and institutions, including those of governance, are the three key variables usually invoked for understanding the growth performance. A glaring feature of the investment scene in the post-reform period is the steady decline in the rate of public investment and a steep rise in the share of private investment. According to the estimates by the CSO, gross capital formation by the public and private sectors was roughly equal at around 10 per cent of GDP each in the early 1980s. But between 2005-06 to 2007-08, public sector investment constituted 8.4 per cent of GDP, while that by the private corporate sector rose to 15.1 per cent and by households to 11.4 per cent (Reserve Bank of India, *Annual Report, 2011-12*).

The decline in public investment is even more glaring at the state level where bulk of the public expenditure on irrigation, power and social sectors is incurred. This is indicated by the fact that the share of the states declined from around 50 per cent of total plan expenditure in the country in the 1980s to 40 per cent towards the close of the 1990s.

Within the states, the per capita plan outlays of the poorer states have always been much lower than those of the better-off states. These disparities have widened in the post-reform period. For instance, during the Sixth Plan period (1980-1985), the actual per capita plan expenditure for the poorest four states, on an average, was a little over half of the average per capita plan expenditure of the better-off states like Gujarat, Maharashtra and Punjab. But during the Ninth Plan period (1997-2002) this proportion came down to around 40 per cent.

Central assistance for state plans (including assistance for externally aided projects), which is a major component of state plan resources, has been progressive in that the per capita assistance for the poorer states has been higher than for the richer states. The poorer states have been handicapped basically by their own weaker resource position. The per capita own plan resources of the poorer states, including market borrowing, constituted around 40 per cent of own per capita plan resources of some of the better-off states, *viz.*, Punjab, Maharashtra, Haryana and Kerala during the Sixth Plan period. This ratio deteriorated to 28 per cent in 2003-04 and further to less than 20 per cent in 2004-05.

An important factor responsible for this deterioration in the financial position of the poorer states is the decline in the tax-GDP ratio of the Centre for quite some time in the post-reform period and the consequent decline in the transfers to the states through devolution as recommended by the Finance Commissions. The loss to the states, that is, the difference between the devolution estimated by the Finance Commissions and the actual devolution, amounts to ₹ 100,000 crore for the decade 1995 to 2005 (Reddy, 2005). The decline in per capita transfers to the poorer states was even greater because the formula for devolution by the Finance Commissions is quite progressive. Therefore, the 12th Finance Commission has done well by raising the tax devolution by 1 percentage point (Government of India, *Report of the Twelfth Finance Commission*, 2004).

As expected, among states, there is a strong positive correlation between the per capita income and tax-GDP ratio. For richer states, because of their higher tax-GDP ratio, their own tax revenues per capita are much higher than those for the poorer states (Singh, 2005).

This is so even when the tax effort of some of the richer states is lower, and that of some of the poorer states higher than the desirable norm considering their per capita income, rate of urbanisation, etc. A multiple regression analysis with per capita own revenues of state as dependent variable and per capita net state domestic product and rate of urbanisation as independent variables shows that Punjab and Maharashtra whose per capita net state domestic product as well as own tax revenues were among the highest during the triennium 2001-2004 could have mobilised 5-10 per cent more revenues, given their level of per capita income and rate of urbanisation. On the other hand, Bihar—the poorest state with lowest per capita tax revenues—showed highest tax effort by mobilising 38 per cent more revenues than could be expected considering its per capita income and rate of urbanisation. Although the tax devolution by the Finance Commissions has been progressive for the last 25 years, this could not offset the weaker resource position of the poorer states so far as resources for plan are concerned. Higher tax devolutions from the Finance Commissions have helped them basically to reduce their dependence on grants-in-aid for meeting their deficits on non-plan revenue account. That is, without the progressive devolution of tax revenue, the Finance Commissions would have had to take greater recourse to grants-in-aid for filling up the non-plan revenue gaps of such states. However, it must be said to the credit of the 12th Finance Commission that they have raised the proportion of grants in the total devolution and earmarked a significant portion of these grants to the poorer states for the development of social sectors like education and health.

The debt-GDP ratios of the poorer states are higher. Because of their lower credit-worthiness, they have not been able to access borrowings from the market to the same extent as the richer states. The per capita market borrowings of the four poorest states which were almost equal to the market borrowings of certain better-off states, *viz.*, Punjab, Maharashtra, Haryana and Kerala during the Sixth Five Year Plan declined to 72 per cent of such borrowings by these states in 2004-05. The inability of the less developed states to access sufficient resources for the development of infrastructure through higher plan outlays has thus emerged as a critical constraint in redressing regional imbalances in development.

Among states, the correlation between Per Capita GSDP growth rate in the post-reform period of 1990s and the index of Social and Economic Infrastructure in 1995 as well as 2000 is positive and significant. Clearly, therefore, the states whose initial or pre-reform conditions were favourable in respect of infrastructure could benefit more from the opportunities opened up, especially in the service sector, by economic reforms and register higher growth rates in GSDP (Rao and Dev, 2003). In the subsequent period, the rank correlation between infrastructure index and per capita income of states was quite high, rising from 0.79 in 1999-2000 to 0.86 and 0.85 in 2007-08 and 2008-09 respectively (Planning Commission, 2012).

Private Investment

Private investment has been higher in high income states where per capita plan outlays have been higher and where, therefore, infrastructure is well developed. For example, according to the IDBI data, the per capita cumulative disbursements by the All India Financial Institutions up to March 2004 were ₹ 15 lakhs for Maharashtra, 14 lakhs for Gujarat and 7 lakhs for Tamil Nadu, as against less than 3 lakhs for Madhya Pradesh, less than 2 lakhs for Uttar Pradesh and less than half a lakh for Bihar. Similarly, in March 2011, investment plus credit-deposit ratios of Scheduled Commercial Banks (as per utilisation) were high for the northern (88%), western (78%) and southern (106%) regions, and low for the eastern (61%), central (57%), and north-eastern (45%) regions. As to the amount of Foreign Direct Investment (FDI) and Foreign Technical Cases (FTC) approved from August 1991 to December 2006, a few advanced states, *viz.*, Maharashtra, Gujarat, Tamil Nadu, Karnataka and Delhi, together accounted for half the share as against the combined share of less than 10 per cent by the four poorest states, *viz.*, Bihar, Madhya Pradesh, Orissa and Uttar Pradesh (Government of India, 2007).

Technical Change

The issue of technical change does not seem to figure prominently in the debate on regional disparities in development in India. This could be partly because much of the technology is embodied in capital equipment and hence is highly correlated with such investment. Also, this may be attributed to the speedy diffusion of frontier technologies like biotechnology and information technology across regions and even among income groups with large differences in asset endowments, when adequate support services are provided. The success of the Green Revolution in India provides a classic illustration of this process. After the initial success with wheat in the North-Western India for a decade beginning from the mid-1960s, it soon spread to different parts of the country by the 1980s covering rice even in the poorer eastern gangetic plains because of its inherent profitability and relatively low investment requirements.

The experience so far with information technology (IT) is equally encouraging and holds the prospect for raising productivity enormously in millions of farms and factories of varying sizes and the government offices throughout the country. This experience underlines the need to exploit the potential offered by these emerging technologies as well as human development for giving a fillip to the catching-up process in the less developed areas of the country.

Institutions and Governance

Historically, the developed states are, in general, characterised by progressive land tenures like the *ryotwari* or the *mahalwari* systems, whereas most of the less developed states were under the exploitative tenures like the *zamindari* and the *jagirdari* systems. Many of these areas were under the princely states for a long period. The social structure evolved under progressive land tenures has been conducive to the growth of enterprise and generated incentives for work, whereas the social structure perpetuated by the exploitative land tenures has been inimical to enterprise and bred apathy.

For understanding the causes of backwardness and for formulating appropriate development strategies, one has to go deeper into the history. For example, for over a century and half before independence, the rate of taxation was very high in the eastern zone and the extraction of surplus was much higher than in many other parts of the country. We know that there has been a significant drain of resources from the country in the colonial period. But there was also a significant inter-regional flow of resources within the country. There was a large net outflow of resources from the eastern region which was partly frittered away in conspicuous consumption and partly invested in other regions in infrastructure like irrigation and transport (Rao, 2005). This is how the less developed regions inherited poor infrastructure and outmoded institutions at the time of independence.

Even after independence, the per capita transfer of resources from the Finance Commission as well as the Planning Commission was lower for many of the less developed states for at lest two decades when the cost of building infrastructure was relatively low (Rao, 2005). The transfer of resources, especially from the Finance Commissions, has become progressive after the 1970s. But in a federal democratic polity, there are limits to such statutory transfers, particularly because many of our 'developed' states too are poor in absolute terms.

The current debate on 'good governance' in India is largely divorced from the nature of the prevailing social structure and the need to generate socioeconomic impulses for its transformation. As such, it comes out essentially with technocratic prescriptions regarding good governance in the poorer states as a prerequisite for attracting greater investment flows and for stepping up growth through the efficient use of resources. It underplays the role of factors like investment in infrastructure, technical change for raising the profitability of investment and making governance accountable to the people by empowering them through the *panchayati raj* institutions (PRIs) in triggering off the desirable changes in the prevailing socioeconomic structure, which is the root cause of poor governance and inefficient delivery systems in the less developed areas.

The Way Ahead

It is clear from the foregoing discussion that for reducing regional disparities in development, improving social and economic infrastructure in the backward regions through greater public investment needs to be given the highest priority in the development strategy. The other two areas of priority action for these regions are speeding up social transformation through the empowerment of the common people and measures for good governance, discussed above and in the first essay on Inclusive Growth. The 12th Plan proposals on improving governance are promising as they centre round the need for empowering people, strengthening the accountability mechanisms, better coordination between various departments and enhancing the capabilities at the grass-roots level, among other things, by involving voluntary agencies.

In this presentation, I have chosen to focus mainly on the inter-state disparities in levels of development. However, growing regional disparities within certain states, especially the bigger ones, is a matter of equal concern. For example, the Twelfth Plan in its chapter on regional equality notes that "though Maharashtra, Karnataka, and Andhra Pradesh are regarded as rapidly developing States, the fact of the matter is that there are a few pockets within these States which, due to their better physical and social infrastructure, have been able to attract large investments and register a faster rate of economic growth. State-level analysis does not reveal anything about what might be taking place in different regions within States. In fact, intra-State disparities are as much a cause of concern as inter-State disparities" (Planning Commission, 2012). It is clear from the development experience and the simmering unrest in several less developed parts of the country that without major initiatives at the national level, the regional disparities in development may accentuate further to crisis proportions threatening social harmony and national integrity, apart from depriving the nation of the opportunities for higher and sustainable development.

Unfortunately, precisely at this moment when imaginative and decisive initiatives are needed towards this end by the Centre, there are conflicting signals emanating from the policy-making levels, and a clear and coherent policy is yet to emerge. The Twelfth Finance Commission, while linking the debt write-off scheme for states with the quantum of reduction in their revenue deficits, decided to fix norms, targets and the time-frame that are uniform across all states regardless of their initial levels of revenue deficit, income levels and debt-GSDP ratios. The permissible level of fiscal deficit under the Fiscal Responsibility and Budget Management Act is uniform across all states. Further, the recommendation of the Twelfth Finance Commission to dispense with the loan component of central assistance for state plans and leaving the states to directly access the market for loans can adversely affect the development of the poorer states whose credit-worthiness is lower, unless, as suggested by the Finance Commission, the Centre volunteers to intermediate effectively on behalf of the poorer states. Therefore, an alternative framework for enabling the poorer states to step up their developmental expenditure needs to be quickly put in place by the Planning Commission.

Certain criteria presently used—whether by the Planning Commission or Finance Commissions—*viz.*, tax effort, fiscal discipline, population control, on-time completion of externally aided projects, etc., need to be revisited. For one thing, the measures of 'tax effort' and 'fiscal discipline' currently being used are far from satisfactory. These are clearly iniquitous because they have an in-built bias towards better-off states. For example, the ratio of own revenue receipts to total revenue expenditure of a state—an important element in the criterion of 'fiscal discipline'—can be expected to be higher for developed states which have been growing faster. In this scheme of things, the advantages of historically driven high growth rates are attributed to greater concurrent efficiency. Besides, the contribution of allocations based on these criteria towards fulfilling the avowed objectives is dubious. This is because several other factors are far more important in improving the indicators chosen.

Mention may be made, in this context, about a large weight (17.5%) given to 'Fiscal Discipline' in the devolution formula used by the Thirteenth Finance Commission by reducing the weight for fiscal capacity distance. As argued above, this, in effect, makes the devolution formula less progressive. This inference is strengthened by the fact that better-off states gained more at the expense of the poorer states from the application of this criterion (in comparison to income distance criterion), when it was assigned a weight of 7.5 per cent by the Twelfth Commission. Of the share assigned for distribution on the basis of this criterion, the six poorer states, *viz.*, Bihar, Chattisgarh, Jharkhand, Madhya Pradesh, Orissa and Uttar Pradesh, have received 37 per cent, whereas they would have been entitled to 62 per cent if the same was distributed on the basis of income distance criterion. On the other hand, four richest states, *viz.*, Gujarat, Maharashtra, Punjab and Haryana received 19.3 per cent as against 2.5 per cent if the same was distributed under the income distance criterion.

The Thirteenth Commission notes that following their recommendations, the ratio of devolution to GSDP would improve for all the states when compared to the corresponding ratio for the period covered by the 12th Commission (Government of India, 2009). However, the improvement noted may be attributable basically to the significant rise in the tax-GDP ratio in the period covered by the Thirteenth Commission. Such an improvement in the ratio of devolution to GSDP would have taken place even if the formula used by the Twelfth Commission were to be applied to the revenues projected by the Thirteenth Commission.

It could be argued that because the loan component of Plan assistance has been dispensing with, there is not much to be gained for poorer states by revisiting the Gadgil-Mukherjee formula. For example, during 2011-12, the share of Normal Central Assistance was reduced to a mere 4.5 per cent of the total Plan transfers to states (Planning Commission, 2012). One should, therefore, look for a more promising source, *viz.*, the pattern of Plan assistance under various centrally funded schemes. Among these, the 14 flagship schemes cover the areas of agriculture, education, health, employment, urban development, rural and urban infrastructure, and energy. A look at the pattern of allocation for such schemes during 2011-12 does not show them to be progressive: the six poorest states, *viz.*, Bihar, Jharkhand, Uttar Pradesh, Madhya Pradesh, Chattisgarh and Orissa, which together account for 40 per cent of population, received 39.75 per cent of total allocation to states (Planning Commission, 2012).

There is, therefore, a strong case for renewing efforts at making the grants component of Plan assistance under Gadgil-Mukherjee formula as well as assistance under various centrally funded schemes more favourable to the less developed states. This is especially necessary if the quantum of grants has to be raised substantially to help such states in view of the constraints in accessing loans in the market. It is indeed ironical that Central assistance for state plans, whose avowed purpose is to ensure speedy development of the poorer states, has been less progressive than the devolution from the Finance Commissions. This could be because the latter has been left to be decided by a few experts, whereas in the case of the former, consensus among states is indispensable (Vithal and Sastry, 2002).

Even so, before making their recommendations, the Finance Commissions, for example, have been holding extensive discussions with state governments as well as with various sections of our population. It speaks highly of the trust that our people have reposed in the constitutional arrangements of our federal democratic polity that the recommendations of the Finance Commissions have been received, by and large, favourably. This suggests that a far-sighted leadership can still generate consensus among states and people at large on the need for a more progressive Formula for Central assistance for state plans in the interests of preserving social harmony and strengthening national integration.

References

Ahluwalia, Montek S. (2000). "Economic Performance of States in Post-Reform Period", *Economic and Political Weekly*, May 6.

Bhattacharya, B.B. and S. Sakthivel (2004). "Regional Growth and Disparity in India: Comparison of Pre- and Post-Reform Decades", *Economic and Political Weekly* 29(10), March 6.

Dev, S.M. and Jos Mooij (2003). "Patterns in Social Sector Expenditure: Pre- and Post-Reform Period" (mimeo), Hyerabad: Centre for Ecnomic and Social Studies.

Dev, S.M. and C. Ravi (2003). "Macroeconomic Scene: Performance and Policies", in C.H.H. Rao and S. Mahendra Dev (eds.), *Andhra Pradesh Development: Economic Reforms and Challenges Ahead*. Hyderabad: Centre for Economic and Social Studies.

Government of India (2004). *Report of the Twelfth Finance Commission* (2005-10). New Delhi: Government of India.

———. (2007). *Handbook of Industrial Policy and Statistics*, 2006-07. New Delhi: Department of Industrial Policy and Promotion, Government of India.

———. (2009). *Report of the Thirteenth Finance Commission* (2010-2015), Vol. I. New Delhi: Government of India.

Planning Commission (2012). *Twelfth Five Year Plan* (2012-2017). New Delhi: Planning Commission, Government of India.

Rao, C.H.H. (2005). *Essays on Development Strategy, Regional Disparities and Centre-State Financial Relations in India*. New Delhi: Academic Foundation.

Rao, C.H.H. and S. Mahendra Dev (2003). "Economic Reforms and Challenges Ahead: An Overview", in C.H.H. Rao and S. Mahendra Dev (eds.), *Andhra Pradesh Development: Economic Reforms and Challenges Ahead*. Hyderabad: Centre for Economic and Social

Studies.

Reddy, G.R. (2005). "Twelfth Finance Commission and Backward States", *Economic and Political Weekly* XLI(32), August 12-18, 2006.

Singh, Ajit Kumar (2005). "Finance Commissions Devolutions and Regional Imbalances", paper presented at the National Seminar on *Accelerated Economic Growth and Regional Balance—Recent Experiences and Implications for Inter-State Variations in Development*, organised by the Indian Economic Association, Institute for Studies in Industrial Development and Institute for Human Development, New Delhi, September 16-18.

Vithal, B.P.R. and M.L. Sastry (2002). *The Gadgil Formula for Allocation of Central Assistance to State Plans*. Hyderabad: Centre for Economic and Social Studies.

Updated version of the Lecture dedicated to the memory of Professor A.M. Khusro, delivered at the 88th Annual Conference of the Indian Economic Association, held at Visakhapatnam, 27-29 December, 2005; published in *The Indian Economic Journal* 54(1), April-June, 2006; Also in Kanchan Chopra and C.H. Hanumantha Rao (ed.), Growth, *Equity, Environment and Population: Economic and Sociological Perspectives*, SAGE, 2008.

II

Agriculture: Growth and Sustainability

9

Crisis in Indian Agriculture

WE have been experiencing agrarian crisis in the country as a whole including Andhra Pradesh (A.P.). There are great similarities between the crisis in A.P. and the rest of the country. But the manifestations of the crisis seem to be more severe in A.P. This is to be expected because the earlier government of A.P. was a votary of particular brand of economic reforms which neglected agriculture. It was so in the country as a whole but it was claimed that A.P. was in the forefront of economic reforms. The government here was very conspicuous in neglecting agriculture and rural development which has brought about this crisis. But what are the manifestations of this agrarian crisis?

The scholars concerned with agricultural development have brought out some of the features of the present agrarian crisis which has been building up for quite some time. In fact, it dates back to the early 1980s or the mid-1980s. It is true that the crisis has become sharper after the introduction of economic reforms in the early 1990s. But so far as agriculture is concerned, the neglect of the sector was becoming evident even during the early 1980s. One could see this in the planning commission in the early and mid-1980s. In fact, while framing the Seventh Five Year Plan in the mid-1980s, some of the discussions centered round the 'neglect' of irrigation. The data which was trickling down from the National Accounts statistics was clearly showing that public investment in agriculture was going down. And, by the late 1980s when the National Accounts were published it became clear that the public investment in agriculture, particularly in irrigation, started declining since the early 1980s. This happened because the focus of agricultural policy was on the Green Revolution in the developed pockets of the country like Western Uttar Pradesh (U.P.), Punjab, Haryana or parts of A.P., parts of Tamil Nadu, etc., which already had irrigation facilities. There was thus a drive to get maximum output of wheat and rice from the already developed areas of the country, where the results were quick and substantial.

So, after 15-20 years of the Green Revolution, the Food Corporation of India (FCI) accumulated enough stocks of food grains to meet the demand from urban areas, from those who have purchasing power. Then the concern for agriculture began to slow down. The concern for backward regions and for small and marginal farmers was already fading away. This was becoming clear even in 1980s. Of course, economic reforms reinforced this trend.

If one looks at the data, particularly in the post-economic reform period, that is, in the 1990s, one finds the first manifestation of this crisis in the declining growth in agricultural output. There is a deceleration or slowing down in the growth of agricultural output, particularly food grains output, in the post-economic reform period of 1990s when compared to 1980s. The second manifestation of this crisis is the declining productivity, i.e., the growth in per acre output has been going down, again more sharply for food crops. But even the productivity of all inputs together—or total factor productivity—which is a good measure of efficiency in agriculture, has also been going down. This happens particularly when there is a neglect of new technologies. The investment in agricultural research and extension was neglected in the post-economic reform period, more so in A.P.

The third manifestation which became acute recently—which in fact led to the fall of previous government, whether in A.P. or in the country as a whole—is the growing rural unemployment. The employment growth in rural areas, particularly in agriculture, went down sharply since the early 1990s, that is, during the economic reform period. This is well known by now and papers by several scholars bring out this fact. The sharp fall in employment growth rate was even more visible and acute in A.P.

The fourth manifestation is the increasing vulnerability of small farmers to weather shocks. These days when one talks of farmers' distress, one gives the impression that it is all because of drought. But if we analyse the data on the incidence of drought or the variability of rainfall, we will see that for the country as a whole or even at the regional level, the variability or the incidence of bad years in the 1990s is not greater than in 1980s or 1970s. You have as many bad years or good years in the earlier decades also. But a bad year now brings much greater distress or hardship than before. This is because the ground water resources have been over-exploited, particularly in the 1980s and the 1990s. The ground water reserves are much smaller now. This coupled with population growth and greater pressure on natural resources leads to greater vulnerability.

Finally, as a result of all these factors, we now find that the farmers' indebtedness has increased, particularly from informal sources, that is, money lenders, traders, etc. The share of institutional sources, that is, banks and co-operative institutions has declined drastically for small and marginal farmers. Not only that there is greater indebtedness but there is greater insolvency because of risk, leading to suicides by farmers.

The basic causes for agrarian distress are common for the country as a whole, including A.P., particularly in the last 10 years, due to the change in economic policies. But, how do the economic reforms affect agriculture adversely? In theory, one can argue that economic reforms need not necessarily harm agriculture. One can get support for this proposition from the experience of China, Vietnam and some other countries. In those countries even after the introduction of economic reforms, their overall conditions and policies were such that agriculture was not damaged. In fact, there was a breakthrough in agricultural output in China immediately after economic reforms were introduced in the late 1970s. Therefore, what is at issue is not the economic reforms *per se*, but the conditions under which they were introduced. The initial conditions were favourable in China; for example, land reforms there were successful, but in India they were a failure. Further, in China there was considerable strengthening of infrastructure like irrigation.

The content of economic reforms in our country was very different so far as agriculture was concerned. There are two broad aspects of economic reforms which effect agriculture. First, the prices of agricultural commodities as a whole, i.e., the overall agricultural price index in relation to non-agricultural prices or industrial prices. In the pre-reform period, Indian industry was heavily protected—in fact our industry was known to be one of the most protected in the world. Economic reforms did reduce this protection to industry. Because of that many of the industrial prices came down relative to agricultural prices. Absolutely all the prices—both industrial and agricultural prices—rose. But the relative prices or what we call terms of trade or the terms of trade index improved for agriculture. This is for the sector as a whole, not for individual commodities. The overall index of agriculture prices in relation to non-agricultural prices improved during this period. It happened in all reforming countries including China.

In response to this, the share of private investment in Indian agriculture increased. But in A.P., surprisingly, even private investment declined. But terms of trade and private investment are a very small component of the benefits flowing to any underdeveloped agriculture. In a developing economy like India, agricultural prospects depend heavily on public investment. Economic reforms as they worked reduced public investment in agriculture. The very philosophy of economic reforms was to reduce public investment, to reduce plan expenditure and to increase private investment in the economy. Reduction in tax rates led to a slow growth of central revenues whether it is income tax or other indirect taxes. Because of that, the share of states in central revenues also declined. All this led to the decline of public investment. This is the second aspect of economic reform affecting agriculture.

Because of the decline in public investment, irrigation continued to be neglected. Today we have nearly 400 major and medium irrigation projects in the pipeline under various stages of construction. They are there for the last 30 or 40 years. If we proceed at the same rate at which we have been proceeding, it will take at least 25 years to complete these projects. When we complete these projects, much of the irrigation potential in the country will have been exploited.

If agricultural research and extension is neglected, there would be no sustaining of the Green Revolution. The Green Revolution in the country owes basically to technological changes—the high yielding varieties (HYVs) of rice and wheat, the application of fertilisers and pesticides, etc. The Indian Council of Agricultural Research (ICAR) has been asking that the government should spend at least 1 per cent of the value of agricultural output for research and extension. Our expenditure has been around 0.3 per cent; for some years, it may be around 0.5 per cent, but basically around 0.3 per cent of the value of agricultural output when compared to 1 per cent projected by ICAR 10 years ago. A.P. is much worse in this respect. Agricultural extension in A.P. became virtually defunct under the previous government. The expenditure on agricultural extension here was around 0.05 per cent of its agriculture gross state domestic product (GSDP), which is very low compared to the level of expenditure incurred at the national level. Because of this, there has been a field day for middle men, that is, private traders, distributors and money lenders in the rural areas. Particularly in respect of credit, there has been a very big change after the reforms. In the 1980s, the target for agriculture was 18 per cent of the total credit from commercial banks. Even this was not realistic. It should have been much higher. But in actual practice, the credit disbursed by commercial banks came down to 11 to 12 per cent or 13 to 14 per cent in some years. Thus the gap was 4 to 5 per cent. To make good this deficiency, the Rural Infrastructure Development Fund (RIDF) has been started through which the National Bank for Agriculture and Rural Development (NABARD) initiates some infrastructure projects in rural areas with a matching grant from the state governments. But, because matching grants were not forthcoming, the utilisation of RIDF came down to 30 per cent or so, and thus 70 per cent of the allocated funds remained unutilised. This went on for years together, and on account of this, even the scheme was discontinued by the earlier government.

But in his budget, after the elections, Mr Chidambaram has revived it without telling us how he proposes to make use of it more efficiently. The new government recognised the rural credit crisis and has promised to double the institutional credit for rural areas in the next three years. And in keeping with that policy, the target is to increase the supply of credit by 30 per cent in each year from the institutional sources. But this is the target for the sector as a whole. We know that the real sufferers are the small and marginal farmers.

They borrow mostly from money lenders or traders. So we need to have a sub-target. It is not sufficient to say that we will double the supply of rural credit from the institutional sources. We know who is going to appropriate this credit. In fact some of the studies in 1970s showed that whenever the supply of institutional credit is liberal, it is the big farmers who borrow at low interest rates and re-lend it to the small and marginal farmers at the going rates of interest which are very high in rural areas. Therefore, if we want to arrest the immiserisation of small and marginal farmers, we need to have a sub-target of credit by stipulating that at least 50 per cent of the increase in institutional credit should go to the small and marginal farmers.

As pointed out earlier, there has been an overexploitation of ground water resources. Area under canal irrigation declined in A.P. in the 1990s. The area irrigated under tanks also declined substantially. In fact thousands of tanks have become defunct in A.P. If the tanks become silted, they do not store enough water. Then the ground water potential also goes down. On the top of it when there is overexploitation by the affluent farmers, who can spend on power and pump sets, then the groundwater table further goes down. And this water is utilised to grow rice and sugarcane which are highly water-consuming. The small farmers too, on seeing some of the affluent farmers becoming rich, try their luck. At least some of them who can obtain sufficient credit go in for such capital-intensive farming which is risky. Then they have to bear higher costs because they have to dig deeper as the water table has already become low. That accounts for the stories regarding failure of wells and suicides of farmers.

In the course of economic development in many of the East and South-East Asian countries, particularly in China, there was a big growth of non-agricultural activities in the rural areas. The rural industries came up in a big way. Because of that the pressure of population on agriculture was eased. Even today in China, more than 60 per cent of the population is dependent on rural areas. But many of them work in rural industries and earn a great deal from the non-agricultural activities in the rural areas. Because of that the pressure on agriculture is low. But here in our country, the development of non-farm activity has been very slow. In the last 10-15 years because of economic reforms, a big boost was expected in non-farming activity in rural areas. But there was no such development in rural industries. Therefore, the pressure on limited land resources continued to be very high. Today, the land per capita held by small farmers is much lower than it used to be 20 years ago. Their asset base has come down both in quantity as well as in quality because the irrigation facilities have been declining. Therefore, when there is drought and there is a fall in output, they become more vulnerable. This is also the reason why there is greater demand for land from the landless people and the small farmers. The area under tenancy is very high in A.P. but it is not recorded. And because of low bargaining power of small and marginal farmers, the rents are also very high. It is surprising to see rental share between 60 per cent and 70 per cent of the total output. Traditionally it used to be anywhere between 25 per cent and 50 per cent. Fifty per cent was the norm in crop sharing. Interest rates on credit also are very high. Thirty-six per cent is the normal rate of interest in rural areas charged by the moneylenders.

One is tempted to ask: Why are the moneylenders lending money to the farmers when they are committing suicide and the lender cannot recover money from them? The fact of the matter is that what the lenders lose because of suicides is not much; it may not exceed 5 to 10 per cent of what they invest. For them, the number of small farmers committing suicides is negligible when compared to millions of farmers borrowing money from them at exorbitant rates of interest. All of them are subjected to the same stress. But the most vulnerable among them are committing suicide. In spite of suicides by the farmers, moneylenders are not only able to recover their capital but are able to make big profits because the rates of interest are very high and exploitative.

It is heartening to note that at least the new government in A.P. has moved quickly to take up new irrigation projects and to complete the on-going ones as early as possible. But, it is not going to be easy. Completing all these projects in the next three to four years requires a very big effort.

One is not sure whether there is clarity on the need for enhancing outlays on research and extension. If one looks at the plan outlays for agriculture in A.P. in the last 5-6 years under the earlier government, the per hectare outlays were the lowest when compared to all the southern states. So this needs to be stepped up. Then only the unscrupulous traders and moneylenders who are selling spurious seeds, fertilisers and pesticides can be held in check. One aspect which needs serious thinking among all those concerned with rural development relates to water issue. Our calculations show that even if all the medium and major irrigation projects are completed, after 20 years or so, we will be facing the same problem of water shortage as we are facing today. The demand is going up rapidly not only for irrigation but also for drinking because of growing population, urbanisation and industrial use. So the problem of shortage of water is going to remain even after exploiting all the irrigation potential by completing all the major and minor irrigation projects. So, how to conserve water, how to use it economically is going to become a major issue. That means choice of appropriate crops, adopting technologies which create less pressure on water and which economise on water use. Free power does lead to wasteful use of water. But it may be argued that because of special circumstances A.P. government is justified in giving free power for some time. However, in the study brought out by the Centre for Economic and Social Studies (*Andhra Pradesh Development: Economic Reforms and Challenges Ahead*), it was suggested that power could be highly subsidised or even given free up to certain units sufficient for 1 or 2 ha. If it is not possible to give free power to only some sections of the farmers, it may be given to all, but free up to only a certain limit. If that is done, 70 to 80 per cent of farmers—small and marginal—will get relief. They account for only 30-35 per cent of the area. The remaining two-third of the cultivated area for which power is used is held by the middle and large farmers who can afford to pay. The government can subsidise them too, but need not give them free.

Even under earlier method when power was charged, one cannot say that power was being economised or water was not being wasted. This is because the charges for power were on a flat rate basis. Under this system, one can pump as much water as one likes. It is a fixed charge. What economists have been asking for is volumetric rate, that is, you charge according to the volume of consumption of electricity.

This is how we can discourage the overexploitation of groundwater. If government fails to do this, communities have to be made aware of the emerging water crisis, that is, water literacy has to be promoted in rural areas and the rural communities and *panchayats* should be fully empowered to manage water resources economically. If the village *panchayat* decides to tax some farmers or some users of water according to the volume of consumption, then the government should not interfere; rather the government should encourage such *panchayats* to economise on the use of water. Rural communities should be encouraged through incentives to charge for extra consumption and to retain that extra income for local development. Some such solution has to be found, otherwise water crisis will be there with us.

The Common Minimum Programme (CMP) promised 100 days of guaranteed employment for one person in each family in a year. The government is going to introduce a bill in the Parliament for Employment Guarantee Scheme. This is to mitigate the rural distress, to help the small farmers and the landless labourers. If this is done some of the pressure on agricultural resources may be eased. Not only that, some of the pressure from the moneylenders and traders on the rural poor may also come down because much of this distress now is due to low bargaining power of the rural poor. If there is employment guarantee even for 100 days per year, the studies show that it will have a favourable impact on the market wage in rural areas. On account of employment guarantee, there is every likelihood of the wage rate in the market going up. It will also have a softening impact on interest rates charged by the moneylenders and on the rents charged by the landlords. So this is one of the effective ways of not only raising the incomes of the poorer people in rural areas but also for improving their socioeconomic position and their bargaining power. This is very necessary to reduce pressure on natural resources.

This paper is based on the inaugural presentation at the Seminar on "Crisis in Agriculture in A.P." held under the auspices of the Sundaraiah Vigyana Kendra, at Hyderabad on 24 July, 2004; published in *Mainstream* XLII(34), April 13, 2005.

10

Suicides by the Farmers

Causes and Remedies

THE cases of suicides by farmers relate almost entirely to the small and marginal farmers and tenants. Studies on farmers' suicides in Andhra Pradesh reveal that such farmers have rarely borrowed from institutional sources like banks and co-operatives. They borrowed almost entirely from moneylenders at exorbitant rates of interest. The foremost disability of such farmers is that their asset base, consisting basically of own land area, has been considerably eroded over a period of time. This is explained by the failure of land reforms to distribute land to the landless and population growth causing sub-division of small landholdings. The own land area per head held by them is much tinier now than a few decades ago. This has not been made good by the improvement in the quality of land through the provision of public irrigation. In the initial years of planning, the small and marginal farmers benefited from the development of irrigation through public sources. Unlike well irrigation, which depends upon the investible capacity of farmers, irrigation through public canals benefits all the classes of farmers in proportion to the area held by them in the command area. Thus the decline in public investment in irrigation in the last two decades is responsible for the substantial erosion of the asset base of small and marginal farmers. They do supplement their holdings by leasing in area, but this does not improve their credit-worthiness, even as they have to pay market-determined rents which adds to their debt burden.

Second, arising in part from the first, the risk of crop failure has considerably increased over time. There is no evidence of any decline in the level of normal rainfall or of an increase in the frequency of droughts. But droughts, when they occur, have become more severe in their consequences, whether it is the non-availability of fodder or of water for irrigation and drinking purposes. A rainfall deficiency of, say, 20 per cent causes now greater hardship than a similar deficiency did a few decades ago. This is because of the very slow growth of public irrigation and the overexploitation of the ground water. Small and marginal farmers, being late comers, face lower water table and thus incur heavy costs for getting water with a greater probability of well failure.

Third, thanks to the Green Revolution and the deficiency of public investment in irrigation, agriculture for an individual farmer has become far more capital-intensive than in the early years of planning. The farmer has now to spend several times more per acre, in real terms, by way of investment in well irrigation and on inputs like fertilisers and pesticides. This has led to the phenomenal increase in the borrowings of small and marginal farmers who do not have investible surplus.

Fourth, the small farmers' distress has been compounded by the virtual withdrawal of public support for institutional credit and research & extension in the post-reform period. There has been a substantial deceleration in the expansion of credit from the scheduled commercial banks to the small farmers in the post-reform period. With hardly any non-farm alternatives available, and under the demonstration effect of the resourceful farmers, the small farmers are showing their willingness to take risk by investing in farming even when their capacity to take risk has been severely eroded. Failure of institutional credit and public extension has provided a fertile ground for unscrupulous traders and moneylenders to lure the gullible farmers to use the spurious inputs in large quantities by extending credit to them.

Finally, because of the deficiency of public investment in rural infrastructure, the growth of rural non-farm enterprises such as agro-processing and other rural industries has been very slow, causing an undue pressure on small holdings. Although the share of agriculture in gross domestic product (GDP) has declined to around 25 per cent, about two-thirds of our population is still dependent on agriculture. In many countries of East Asia, particularly in China, the growth of rural industries in the post-reform period has led to the rise in off-farm employment and wages, thus reducing pressure on small holdings. In

India, where nearly 75 per cent of farmers are still small and marginal, growth of non-farm rural sector alone can provide the much needed safety net for them.

What all this implies is that, from a long-term angle, a big push to public investment in rural infrastructure, including irrigation, research and extension in the drought-prone areas is essential for making farming viable for small holders and for facilitating the development of alternative sources of livelihood for them like rural non-farm enterprises.

In the short-run, much needs to be done to instill confidence among the vulnerable groups of farmers. Moratorium of debt and its rescheduling is essential to alleviate the misery of small and marginal farmers. After all, such measures were frequently resorted to even in the colonial era. But to ensure that the credit flow to the small farmers is not adversely affected, there has to be a big drive immediately through the NABARD for the supply of institutional credit to such farmers by adopting innovative methods.

The big subsidy on borewells for the small and marginal farmers already announced by the state government needs to be institutionalised by launching a corporation for drilling wells for these farmers. This can be combined with insurance for borewells by subsidising the premia. There has been a total breakdown of agricultural extension in the recent past. There is, therefore, an urgent need to revive it, especially for the benefit of small and marginal farmers so as to take them out of the stranglehold of the moneylenders and traders. Such public support alone constitutes real insurance to the farmers complementing the conventional crop insurance under implementation. Community effort at the village level is essential to identify the vulnerable farmers and instill confidence in them by bringing the benefits of government programmes to their doorstep.

The Hindu, June 11, 2004.

11

What Constrains Agricultural Growth in India?

THE deceleration of agricultural growth in India after the mid-1990s is explained basically by the supply side constraints arising from squeezing of public investment in irrigation, soil and moisture conservation, agricultural research & extension and rural infrastructure in general. Sluggish growth in demand for farm products, especially for non-cereal food, could emerge as a major constraint if rural employment does not pick up. But if public investment in rural infrastructure is stepped up, the ensuing diversification of agriculture and growth of rural non-farm enterprises would significantly increase the employment opportunities in rural areas.

That demand could not have been a primary constraint is illustrated by the deceleration in the output growth for wheat and particularly for rice. In both the cases, even though per capita consumption has been declining, minimum support prices were operative at a high level resulting in the accumulation of embarrassingly large stocks which were off-loaded for exports at highly subsidised rates. Evidently, failure to augment water resources and evolve cost-reducing technologies led to a decline in profitability and growth. The downward trend in world prices after the mid-1990s did have some dampening effect on Indian agriculture, especially in respect of oilseeds which experienced a sharp deceleration in growth. This could have been mitigated through timely action within the World Trade Organisation (WTO) commitments. However, in the globalised scenario an efficient solution, in the long-run, is to make Indian agriculture competitive through infrastructure development, specialisation according to comparative advantage and adoption of cost-reducing technologies.

The on-going diversification of agriculture, among others, into dairying, poultry, fisheries, meat, vegetables and fruits is going to be a major source of farm growth for quite some time. This is primarily on account of the shift in domestic demand from cereals to these food items. The per capita consumption of cereals declined significantly since the early 1970s both in rural and urban areas. There has been a marginal increase in per capita consumption of cereals among the bottom 30 per cent of population in rural areas between 1973-74 and 1999-2000. The top 30 per cent of population, despite the necessary purchasing power, showed a decline of around 25 per cent in their per capita consumption of cereals. The decline is even steeper—33 per cent—among the top 10 per cent of population. Obviously, the decline in per capita consumption of cereals is attributable to changes in tastes and preferences. In 1999-2000, the difference in per capita consumption of cereals between the bottom and top 30 per cent of population in rural areas was only about 20 per cent, while there was virtually no difference in urban areas.

On the other hand, the difference in per capita consumption of non-cereal food between these two groups in rural areas in 1999-2000 was as high as 250 per cent. The difference was even greater in urban areas. Clearly, raising the purchasing power of the poor and the middle income groups through employment generation holds the key to tapping this potential demand for non-cereal food which is becoming a major source of farm growth.

Public and private investment in rural infrastructure to ease supply constraints would, by itself, generate significant employment opportunities. Besides, employment elasticity is much greater in the production of non-cereal food, their processing and marketing.

On account of opening up the economy and generally favourable terms of trade for agriculture, private investment in agriculture rose significantly—by about 50 per cent—between 1993-94 and 2003-04, whereas there was, in general, a stagnation in public investment. In the wake of diversification, agriculture is becoming more capital-intensive and risk prone. Therefore, price policy for wheat and rice, while providing minimum necessary support, has to be made conducive to crop diversification. Rise in public investment in rural infrastructure with adequate supply of institutional credit would stimulate private investment further. In view of higher investment requirements, high risk, the need for improved management practices and higher profitability, public-private partnership has become essential, especially in rainfed agriculture. It is also necessary to encourage contract farming between the farmers, including the small farmers, and the corporate sector. Such arrangements can reduce price uncertainty for farmers while improving their access to credit, technology, inputs and markets.

Economic Times, September, 27, 2005.

12

Agriculture and the Union Budget

The Union Budget for 2006-07 provides an excellent opportunity for examining how the long-term problems faced by Indian agriculture need to be addressed by the Budget. There are four major problems confronting Indian agriculture which needed to be addressed by the current Union Budget. These are: (1) growing scarcity of water for irrigation due to the decline in public investment for over two decades; (2) decline in the growth of crop yields owing to technology fatigue, especially in dry land areas, and disruption in extension services because of the decline in public support; (3) unbalanced use of fertilisers, i.e., adverse NPK ratio, due to the rise in subsidy for nitrogen as well as to the rise in the cost of phosphatic and potassic fertilisers following their decontrol, leading to the depletion of soil nutrients, degradation of environment and decline in the productivity of fertilisers; and (4) inadequate access to credit for small and marginal farmers from the institutional sources, i.e., from the commercial banks and co-operatives. The issues arising from these four major concerns and how the current Union Budget for 2006-07 seeks to address them are discussed in this note.

Irrigation

According to the Common Minimum Programme (CMP) of the United Progressive Alliance (UPA) government, "irrigation will receive the highest investment priority and all on-going projects will be completed according to strict time schedule". Resource poor states, *viz.*, Uttar Pradesh, Bihar, Madhya Pradesh and Orissa, account for as much as 55 per cent of the remaining irrigation potential of about 20 million ha from the major and medium irrigation projects. Even to complete them in 10-15 years, the rate of investment will have to be doubled.

Under the accelerated Irrigation Benefit Programme (AIBP), the Centre has been assisting the state governments for the completion of truly last mile projects. Projects approved under AIBP would require balance investment of ₹ 50,000 crore—as against ₹ 15,000 crore provided so far—for bringing in 8 million ha under irrigation.

Allocation for AIBP in 2005-06 was ₹ 4500 crore. The allocation of ₹ 7121 crore for the AIBP in the current budget is about 500 crore less than the amount projected earlier for achieving this year's irrigation target. The Finance Minister has taken credit for adding 6 lakh ha, but this is hardly encouraging when seen against the target of additional 1 crore ha to be brought under irrigation between 2005-09 under Bharat Nirman. Besides, one does not get any idea as to how much of the current allocation under the AIBP would benefit the less developed and resource-poor states where bulk of the irrigation potential remaining to be tapped is located.

The Finance Minister has announced a programme for repair, renovation and restoration of 20,000 water bodies with a command area of 1.47 million ha at a cost of ₹ 4481 crore with the participation of states and with possible funding from multi-lateral agencies. This is a promising initiative and needs to be pursued vigorously as it would significantly benefit the small and marginal farmers for whom such water bodies are an important source of irrigation.

Agricultural Research and Extension

Central government expenditure on agricultural research and education , plan plus non-plan, in 2005-06 (revised) was ₹ 1900 crore, which was 42 crore less than the budgeted amount. This constituted less than 0.3 per cent of agricultural gross domestic product (GDP). Together with expenditure by states, the expenditure on agricultural research and education constitutes hardly 0.5 per cent of agricultural GDP, as against 1 per cent of agricultural GDP projected by the Indian Council of Agricultural Research nearly 10 years ago, keeping in view our requirements as well as the outlays in many of the developing countries.

For 2006-07, the initiatives announced by the Finance Minister on National Agricultural Research with international collaboration, an allocation of ₹ 100 crore for Punjab Agricultural University and elevating the Rajiv Gandhi Centre of Biotechnology, Thiruvanathapuram, to the status of an autonomous national institute are laudable but do not amount to a total picture regarding the priority accorded to agricultural research, education and extension. The budgetary allocation for agricultural research and education for 2006-07 is ₹ 2160 crore—a rise of a mere 10 per cent and substantially less than the requirement projected. Further, the allocation of 245 crore for agricultural extension, which has been in shambles, is also far from adequate. However, the Central Horticultural initiative and the provision of ₹ 1000 crore for food processing sector are encouraging.

Fertiliser Subsidy

Total subsidy on fertilisers in 2005-06 at about ₹ 17,000 crore turned out to be around ₹ 1000 crore more than the amount budgeted. The subsidy on fertilisers now is at least 8 times more than the expenditure on agricultural research and extension and over twice the total expenditure on agriculture. These subsidies have led to a highly unbalanced use of fertilisers with excessive use of nitrogen in relation to phosphatic and potassic fertilisers. This has resulted in lower productivity and degradation of land and water resources. Reduction in subsidies on nitrogenous fertilisers would, therefore, increase crop yields through the balanced use of fertilisers and would contribute to sustainable use of land and water. Targeting fertiliser subsidies to the small and marginal farmers who constitute 80 per cent of holdings but account for only about one-third of operated area would result in substantial savings. A reduction of this subsidy even by 25 per cent, through targeting to the deserving farmers, would enable the government to raise the expenditure on agriculture, including agricultural research and education, by 50 per cent which would in fact raise the demand for fertilisers. If such targeting to small farmers becomes politically and administratively infeasible, the same purpose could be served substantially by limiting subsidy to 1 or 2 ha for all the farmers.

The Finance Minister, instead of coming out with a politically feasible formula—however modest—for reducing fertiliser subsidies, has chosen to postpone the issue by appealing for a consensus on the issue!

Access to Institutional Credit

The UPA government promised that it would double the supply of institutional credit to agriculture in three years. Accordingly, the annual targets set for credit as well as the number of new farmers financed are being achieved. Also, all the commercial banks have been advised to charge not more than 9 per cent interest on crop loans up to ₹ 50,000. However, more assessment is needed on certain qualitative aspects bearing on the adequacy of credit flow, type of beneficiaries, credit delivery mechanisms, actual interest rates charged, etc.

For example, as against the target of 18 per cent of net bank credit to agriculture, the actual percentage has been going down. Is there any reversal of this trend now? Secondly, the amount of institutional credit flowing to small and marginal farmers is reported to be 32 per cent as against their share of 36 per cent in operated area. Besides, there is a need to replace the informal sources which meet over three-fourth of total credit requirements of such farmers by charging very high rates of interest. Thirdly, we do not have any information about innovative methods of credit delivery, if any, such as group-lending to farmers, lending through genuine and bonafide NGOs, etc. Fourth, the amount of loans granted to tenants and sharecroppers in 2004-05 is reported to be only around 360 crore or 1.65 per cent of total lending to new farmers. Similarly, Eastern and North-Eastern region accounted for a mere 6.67 of credit flow. We need to know the constraints faced in lending to such farmers and regions.

In the light of this, the Finance Minister's proposal in the current budget for opening a separate credit window for self-help groups of tenant farmers with a view to extending to them certain proportion of total credit is well taken. But what about the volume of credit flowing to small and marginal farmers in general? Instead of addressing this issue of credit delivery seriously, the Finance Minister announced the appointment of another expert committee on this.

The Finance Minister has provided a sum of ₹ 1700 crore as relief to the farmers on their interest liability upto 2 percentage points on their borrowings from institutional sources upto ₹ 100,000 during 2005-06. This should have been targeted entirely at the small and marginal farmers. This absence of targeting is glaring in his announcement of lower interest rate of 7 per cent for short-term credit, with an upper limit of ₹ 3 lakh on the principal amount, with effect from Kharif 2006. Even for small farmers, the real need is the adequate availability of institutional credit, as they have to borrow from the informal sources at interest rates often exceeding 60 per cent. Unless credit delivery systems for the poor are attended to and a reasonable proportion of credit pre-empted for them, the benefit of lower interest rate may accrue to the affluent sections who may re-lend to the poor at very high interest rates.

Finally, while the proposal to raise the credit under the Rural Infrastructure Development Fund (RIDF) from ₹ 7300 crore to ₹ 10000 crore in the current budget is welcome, the Finance Minister should have indicated the steps for improving its utilisation, which is reported to be less than 60 per cent, despite the fact that these advances are not entirely for agriculture.

Presented at the Seminar on "The Union Budget for 2006-07" organised by the All India Trade Union Congress in Delhi; published in *Mainstream*, March 17-23, 2006.

13

Biotechnology for Indian Agriculture

Potential, Performance and Concerns

TO start with, the good news is that, of late, agriculture is back on the agenda of the policy makers in India after a pause of nearly two decades. Although the current crisis in Indian agriculture is not comparable to the one faced in the mid-1960s, the prospects for sustaining food security are being seriously called into question if we proceed with business as usual.

This is because area under cultivation may, in all probability, decline significantly in the next few decades on account of demands from growing industrialisation, urbanisation, housing and other non-agricultural uses. For similar reasons, there would be increasing pressure on the already scarce water resources. Despite some success stories of the last decade such as that of Bt cotton, Indian agriculture has, by and large, been experiencing a technology fatigue. This is evidenced by the deceleration in the total factor productivity (TFP) growth in agriculture, as technological changes are a major determinant of TFP growth. It is clear by now that unless there is a breakthrough in TFP growth, the country may be driven to extreme dependence on imports in respect of vital farm product in the next two decades or so.

In retrospect, the controversy surrounding Green Revolution technology introduced in the mid-1960s to overcome food shortages appears much less vociferous than that surrounding biotechnology now. The basic dissent then came from the structural school, mainly confined to a group of social scientists in the academia, according to which unless land relations are drastically changed, it would not be possible to raise agricultural productivity. This view evaporated very soon with substantial increases in yields achieved even by the small and marginal farmers adopting new technology.

The criticism that gained credibility then, and applicable even now to biotechnology, concerns the need to evolve technologies for the less favourable areas and for strengthening the infrastructure and resource position of the less developed regions and small and marginal farmers, so that the adoption of new technology is sufficiently broad-based with an equitable sharing of gains. The degradation of environment attributed to Green Revolution technology seems largely misplaced, as it is difficult to trace this to the nature of technology as such, but is attributable mainly to faulty socioeconomic policies and institutional arrangements resulting in over-extraction of groundwater and excessive and unbalanced use of fertilisers.

The three major issues of concern with respect to biotechnology are: safety for the health of animals and human beings, uncertainty and risk of yield outcomes, especially in the context of heavy investment requirements, and adverse impact on environment by affecting biodiversity. These are basically issues of science and so need to be pursued at that level. The social scientists and social activists can, however, identify, highlight and communicate the relevant concerns to the community of scientists, so that they can be addressed on a priority basis.

Farmers in India, as elsewhere, have, by and large, shown a positive approach to biotechnologies, which indeed explains the widespread adoption of Bt cotton in the country, and also explains why, on all accounts, the suicide rates for cotton farmers are much lower now than in the pre-Bt technology phase. The civil society activists seem to be, on the whole, highly skeptical of biotechnologies while social scientists, including economists, have not evinced adequate interest in this vital area.

The present atmosphere of confusion and skepticism is not conducive to realising the full potential of biotechnology for increasing farm productivity. There is an immediate need for a two-tier institutional mechanism for clearing the air and speeding up the rate of technological advance. First, a high-level forum put in place for dialogue between agricultural scientists, social scientists and officials concerned with the diffusion of biotechnology can help the scientists in setting up an appropriate research agenda, even as it can promote a better understanding of the relevant issues among the social scientists and the administrators. At the next level, there can be an extended body comprising additionally representatives of civil society, farmers, consumers and the media for the exchange of views.

At the public policy level, there are some heartening signals in the present biotechnology era such as the prospect of higher public investment in agricultural research and development (R&D) in the 12th Five Year plan. But the position regarding institutional arrangements for decentralisation and autonomy of public research institutions is not clear. There is, by now, enough clarity on these issues, but lack of will for action at the highest level is holding up the needed reforms. It is clear that without such reforms no breakthrough in research is possible.

Biotechnology opens up wide-ranging possibilities for raising productivity of crops and animals and lending stability to yields. But such prospects also bring out clearly the inherent risks for health and environment from certain innovations. These fears gain credence because of increasing role of private sector research and extension that is commercially driven with a shorter time-horizon. It is, therefore, essential that public sector research and extension is strengthened in this area with appropriate regulatory framework and institutional arrangements for better communication between various stakeholders.

Although private R&D and extension is becoming increasingly important, a high complementarity continues to exist between public and private endeavours. The intellectual property rights regimes and other regulatory framework have to be such as not to constrain the role of public policy in addressing sensitive issues of public concern even as they should not adversely affect the incentives for the private sector effort.

Opening Remarks at the International Seminar held at the Centre for Economic and Social Studies (CESS), Hyderabad, January 18, 2012.

14

Patterns of Agricultural Growth and Links with Nutrition

Agriculture-Nutrition Linkages

Agricultural growth rate well above population growth rate through the rise in productivity of inputs would result in a decline in the relative prices of food. This would ensure availability of adequate quantity of food at affordable prices. Second, pro-poor growth would generate adequate income and purchasing power for small and marginal farmers and wage labour among whom malnutrition is pervasive. Third, gender-sensitive growth with effective participation of women farmers in decision-making by exercising control over income and expenditure can ensure adequate nutrition for them and for their children. Reduced work burden at home and in the field by lowering demands on physical energy can improve their health status. Fourth, diversification of agriculture and rising share of high value food—rich in micro-nutrients—can improve access to such food for farm households because of their ready availability as well as securing higher income. Bio-fortification of food crops with micro-nutrients can further enhance their nutritional value. Finally, agricultural growth high enough to stimulate and sustain non-farm rural growth through its forward and backward linkages would result in higher wages for farm labour as well as greater off-farm employment for the poor with higher and more stable income.

Recent Experience

According to TANDI (Tackling the Agriculture-Nutrition Disconnect in India), the impact of accelerated per capita income growth on improvement in nutrition in India in the recent period is much weaker than that suggested by global experience. Slow growth of agriculture and consequent reversal of the decline in the relative prices of food grains could be a significant factor accounting for this disconnect. The weakening of public support systems for research, extension and institutional credit may have caused greater slow down in the income growth for small farmers and wage labour who predominate among the malnourished. Owing to population growth and sub-division of land holdings, land per head under cultivation has been declining. Pressure on agriculture has increased owing to slow growth of output and employment in the rural non-farm sector.

Agriculture has not fully benefited from the demographic dividend owing to the increasing out-migration of adult male labour. The consequent feminisation of agricultural labour force in the absence of property rights on land for women, their low literacy levels and the absence of gender-sensitive public support systems for extension and credit may have resulted in low productivity and income while at the same time increasing the work burden on the already undernourished women farmers.

That agricultural growth has been skewed against the rural poor in India in the recent period is evident from the rise in the incidence of malnutrition among them. Estimates based on the National Family Health Survey (NFHS) data show that between 1998-99 and 2005-06, the incidence of malnutrition (underweight) among below 3 years olds has increased by 1-3 percentage points for the bottom three deciles (with respect to the standard of living index) as against a decline by 3-7 percentage points for the top three deciles. These top deciles also show a significant decline in severe malnutrition to the extent of 4-7 percentage points whereas there is little change for the bottom three deciles. Among rural women, aged 15-49 years, the incidence of chronic energy deficiency (CED) increased by 1-3 percentage points for bottom three deciles over the same period, while it declined by 9-16 percentage points among the top three deciles.

Prospects

The on-going phase of agricultural growth offers a great potential for attacking malnutrition through diversification of agriculture, strengthening the position of women farmers and accelerating rural non-farm growth.

Diversification of Agriculture

Diversification of agriculture, underway, into micro-nutrient-rich and high-value products like dairying, horticulture and fishery, etc., holds considerable prospects for ensuring nutritional security. Being land-saving and labour-using, these activities typically suit the small farmers. However, being input-intensive, perishable and risky, they need strong support for extension of technology, credit, insurance and marketing services. Small farmers need to be brought together into cooperatives for reducing transaction costs by entering into contracts with the agro-processing firms which could also provide extension services and credit. A big push for public investment in irrigation, agricultural research and extension and rural roads needs to be supplemented by private investments in storage and other marketing infrastructure.

Making Public Support Systems Gender-Sensitive

Diversification of agriculture suits women farmers, as they are traditionally engaged in these enterprises. Provision of institutional credit needs to be based on the viability of production plans and repayment performance, and de-linked with titles to land. The outstanding performance of women's self-help groups in India with respect to bank-linked credit and micro enterprises amply demonstrates its feasibility.

To meet the needs of feminising agriculture, extension services need to become gender-sensitive and feminised by inducting women functionaries in large numbers at various levels in the services chain including agricultural research stations. There is an immediate need for the provision of technologies for women farmers to reduce drudgery from physical labour at home and in the field. Their literacy levels, knowledge and skills need to be improved to meet the needs of diversified agriculture.

Accelerating Rural Non-Farm Sector Growth

Rural non-farm sector has a considerable potential for growth, as it now accounts for only about a half of output and less than one-third of employment in the rural sector. Diversification of agriculture and rise in farm incomes offer potential for the growth of agro-processing and other rural industries. Public investment needs to be stepped up in rural infrastructure, like roads, power, education and skills, and institutional credit. Farmers with small holdings and landless labourers may increasingly shift to the emerging rural occupations both because of higher wages and more stable employment. These would have a favourable impact on their nutritional status.

Keynote Address on *Patterns of Agricultural Growth and Links with Nutrition*, at the International Conference on "Leveraging Agriculture for Improving Nutrition and Health", International Food Policy Research Institute (IFPRI) , in New Delhi, February 11, 2010; Published in *Economic Times*, December 24, 2011.

15

Agricultural Pricing in India

AGRICULTURAL pricing policy has become a highly sensitive subject in the recent period. This is because pricing policy affects different interest groups—farmers, manufacturers and consumers at large. There are, therefore, various points of view or ways of looking at the problem. There is a farmers' point of view, rather, the point of view of those who stand to gain most or lose most from price movements. These are essentially large farmers who sell a substantial proportion of their output. There is the point of view of industrialists who stand to gain from 'lower' prices for agricultural produce, particularly raw materials in the short-run. Again, there is a consumer angle favouring lower prices of food grains. Political opinion at any point of time reflects all these angles, and ultimately it is the reconciliation of the interests of different groups that results in the fixation of particular prices.

In the fixation of such prices, the point of view of experts plays a major role. Although the expert opinion is by no means unanimous, it is still very different from the opinions held by interest groups in so far as the expert opinion is based upon the data on agricultural production and costs collected and analysed in a scientific manner. In view of the sensitive nature of the subject, it becomes extremely important, though very difficult, to approach the subject of agricultural pricing dispassionately from the point of view of agricultural development, overall economic growth and social justice. The Commission on Agricultural Costs and Prices is engaged precisely in such exercises and gives its recommendations to the Government on the prices to be fixed for different crops.

For evolving a sound agricultural pricing policy, it is necessary to bear in mind the basic objectives and functions of agricultural prices. Efficient allocation of resources can be regarded as one of the major functions of agricultural prices. The prices of various crops should normally reflect the relative scarcities of individual crops. Theoretically, the ideal set of such prices of individual crops or equilibrium price structure is the one at which the supplies of the outputs of individual crops are equal to the demands for them. In practice, however, there are often imbalances between various supplies and demands of crops and, therefore, the pricing policy should be such as to induce the output of those crops which are in short supply, in relation to those which are relatively abundant. Relative prices assume special significance for agriculture, because the output of individual crops is highly responsive to the changes in prices. This high price elasticity of supply is explained by the fact that agricultural resources, e.g., land, irrigation, labour, draught power, fertiliser, etc., are easily transferable from one crop to another, though with a certain time-lag.

Pricing policy in the last few years has been favourable to wheat and rice, in the case of which public procurement operations have served to ensure remunerative prices to farmers. This has resulted in the building up of adequate stocks of these commodities. The performance in the case of oilseeds and pulses, on the other hand, has not kept pace with that of major cereals. Although technological advantage for these cereals is a major factor explaining this disparity in performance, relative attractiveness of prices and effectiveness in procurement are by no means unimportant in explaining this phenomenon. For example, in the case of groundnut, the price-spread or the difference between the price paid by the consumer and the price received by the producer is much larger than in the case of wheat and rice.

The second major function of prices is to ensure the desired rate of growth of agricultural output. The allocative and growth functions are, of course, interrelated . Whereas the allocational objective focuses on the relative price structure, the growth objective focuses on the overall price level and the rate of return to investment in agriculture as a whole. Agriculture in India is becoming increasingly a commercial or business proposition. Purchased inputs such as pump sets, fertilisers, pesticides, etc., are becoming important in the overall cost of production. The pricing policy for these inputs and outputs has to be such as to ensure a reasonable rate of return on investment made. Otherwise, the farmers may save less by consuming more or may divert their savings for investment outside agriculture, including the residential construction.

However, while fixing agricultural prices, the rate of return or profit to be allowed should be such as not to reduce the profitability for industries using agricultural raw materials. The cost of wage-goods has also to be kept in mind. There is, thus, a need to strike a balance between the interests of agricultural producers, manufacturers and workers with a view to ensuring a satisfactory rate of overall economic growth.

Equity or social justice is the third important objective of agricultural pricing. We have a large population which is not able to meet the minimum subsistence needs. The proportion of income spent on food is quite high for such population and, as their income level increases, the proportion of household budget allocated for food declines. Therefore, the price level for food grains has an important bearing on the poverty level. Inequalities in income is another dimension of equity. Because of the unequal distribution of land holdings, the proportion of agricultural output marketed to the output produced increases sharply with the increase in the size of holding. In view of this, unduly higher prices for agriculture produce would result in accentuating income inequalities between different classes of farmers.

Higher prices for food grains may cause an increase in inequalities in income for the community as a whole, because the real incomes of the poorer sections of the population will decline at a higher rate than those of the upper income groups for any given rise in the prices of food grains. It has, therefore, become necessary to supply food grains to the vulnerable sections of the population at reasonable prices. However, the subsidies on food grains have been mounting because of the pressures to give higher prices to farmers and charge lower prices to the consumers. All these subsidies have to be borne by the community as a whole in the form of taxes paid or through deficit financing, which ultimately results in raising the overall price level.

It is, therefore, clear that agricultural pricing is essentially an exercise in balancing or reconciling the conflicting objectives cutting across sectional interests, *viz.*, farmer interests, manufacturer interests, consumer interests, etc.

Whereas allocative efficiency, growth and equity are the three major objectives of pricing policy, the latter is only one of the means for achieving these objectives. Public investment in infrastructure such as irrigation, introduction of new technology and institutional changes, such as land and credit reform, are some of the other major policy instruments for achieving the above objectives. Price policy alone cannot achieve these objectives without the judicious mixture of other instruments. If there is no increase in inputs like public irrigation, if there are no improvements in technology and if the supply of credit is limited, increase in prices of agricultural produce cannot by itself bring forth additional supplies because of the non-availability or inelastic supply of inputs. In such a situation, increases in agricultural prices will only result in increased money incomes for those affluent farmers who have substantial surplus to sell. This would accentuate income inequalities in the agricultural sector and in the economy as a whole.

Public investment and efforts towards the provision of essential physical and institutional infrastructure may result in lowering the market prices of agricultural produce because of the significant increases in output. In the case of food grains, such a decline in prices may improve income distribution in real terms. Besides, the provision of public infrastructure distributes benefits of growth equitably because of equal access to inputs for different classes of farmers.

Pricing of agricultural commodities by the Government is usually associated with controls in varying degrees. Agricultural prices are fixed by the Government both in periods of shortages as well as when supplies are plentiful. Price controls, producer levies, rationing, etc., become necessary in periods of scarcity in order to protect the vulnerable sections of consumers. In periods of adequate or plentiful supplies, pricing and procurement becomes necessary for protecting the interests of farmers. This is because the demand for essential commodities like food grains is relatively inelastic and a rise in output causes a more than proportionate decline in their prices, so the cash incomes of the farmers tend to decline despite the rise in output.

In the course of the last 50 years, we have experienced periods of shortages as well as periods of relatively good harvests. Pricing policy, procurement and distribution operations have been geared to meet these two types of situations. In periods of shortages, a good part of the marketable surplus of food grains has been procured at prices fixed by the Government which were lower than the free market prices. However, the prices fixed did cover the cost of production and left some margin of profit. The farmers were free to sell the remaining part of their marketable surplus—usually more than half of the surplus—in the free market at higher prices. It is important to note that procurement of food grains by the Government, in so far as it reduces supplies in the market, pushes up the free market prices. Therefore, in assessing the profitability for the farmers, one has to take into account the weighted average of the prices fixed by the Government as well as those ruling in the market.

Such a 'dual pricing' system has many advantages. First, it enables the Government to procure a reasonable quantity of food grains at prices fair to both producers and consumers. Secondly, it enables the farmers to benefit from higher prices obtaining in the market. This would induce the farmers to make necessary investments in agriculture for increasing agricultural output.

The experience with agricultural pricing policy in India shows that it has, on the whole, contributed to raising the agricultural output by providing incentives to the farmers. The breakthrough in output in the case of wheat and rice is attributable, in a considerable measure, to the policy of price support for these commodities. The growth of these commodities has in fact been achieved at the expense of pulses and oilseeds to some extent. In the absence of public procurement at remunerative prices, the market prices of these commodities would have crashed in quite a few years leading to a slower rate of growth of output. The adoption of new technology as well as investment has been very much facilitated by the pricing policy followed.

Economic Development in India (EDI), Vol. 120, Academic Foundation, New Delhi, 2007.

16

Dual Pricing for Food Grains

IT is reported that the Government is close to finalising a system of dual pricing for the public procurement of food grains. A fixed minimum support price (MSP) covering the cost of cultivation, as at present, recommended by the Commission on Agricultural Costs and Prices (CACP), and variable procurement prices, at the discretion of the Department of Food, depending on the market conditions in different seasons and regions of the country, are reported to be the two basic elements of the proposed system. Under these, the Government would procure quantities enough for meeting the requirements for buffer stocks as well as the Public Distribution System (PDS), in competition with private trade.

If true, this would represent a basic departure from the price and procurement policy followed for nearly four decades since the launching of the Green Revolution. Though belated, this move should be welcomed, as it would enable the country to more effectively cope with the emerging challenges in agriculture. This proposal also recognises the greater ability of our economy, in the post-liberalisation era, to meet temporary shortages of food grains through imports.

One of the major challenges facing Indian agriculture is the slow pace of its diversification on account of an open-ended procurement of food grains at prices much higher than those recommended by the CACP. The food basket, on the other hand, has become considerably diversified with much greater share being occupied now by non-food grain items such as sugar, edible oils, milk, meat, eggs, fish, vegetables and fruits. These items now account for over 60 per cent of consumer expenditure on food. Even for the poor or bottom 30 per cent of population, these non-food grain items of food account for as much as half the total expenditure on food. Since raising the consumption of such items by the poor to the level enjoyed by the top 30 per cent of population has to be the ultimate goal, the per capita consumption of these items by the poor has to increase at least threefold.

By guaranteeing MSP to farmers, the proposed dual pricing system for food grains can safeguard the food grain security already achieved, even as it would encourage diversification of agriculture. Diversification will ensure effective food security by raising employment and purchasing power for the rural poor, as dairying, horticulture, etc., have a potential for greater employment as well as higher wages.

The existing price and procurement policy has encouraged regional concentration of food grains output, with north-western region alone accounting for the bulk of the surpluses procured. This is already causing severe ecological degradation in this region with the depletion of soil nutrients and water table driven down to unsustainably low levels. Open-ended procurement at higher prices without any regard for the emerging trend in international prices rendered our grain exports uncompetitive for quite some time and also reduced domestic consumption by making food grains unaffordable by the poor.

For example, according to a working group of the Planning Commission, over the 5-year period ending 2001-02, when there was a steep rise in procurement prices resulting in the accumulation of embarrassingly large stocks of food grains of over 60 million tonnes with the Government, consumption of food grains in the country was reduced, on this account, by at least 5 million tonnes per annum. The new system by assigning a greater role for private trade can improve the efficiency in the distribution of food grains and substantially cut down subsidies which can help to step up much needed public investment in agriculture.

The proposed dual pricing system is a better alternative than disbanding altogether public procurement at MSP. Such a dismantling could lead to a crash in market prices of food grains in years of good harvests. Even though this may make exports competitive and raise domestic consumption of food grains in the short-run, it may undermine food security by sapping the producer incentives. The experience of the Green Revolution underlines the importance of assured minimum support prices in inducing the farmers to step up their investment and effort and derive full benefit from the available infrastructure.

For the dual pricing system to yield the desired results, it needs to be backed by several other policy measures. Since the impetus for crop diversification would be greater in the infrastructurally developed regions like the norh-west, this can slow down the growth of food grains output in the country and, in particular, the surpluses procured, unless immediate measures are taken to strengthen public support for irrigation, technology, extension and credit in the rest of the country, especially in the central and eastern regions where there is a large potential for increase in food grains output. This is essential as the growth of food grains output has barely kept pace with population growth since the mid-1990s.

This also calls for more effective public procurement of food grains at MSP in these regions, as farmers have often to sell their produce immediately after the harvest at prices that are lower than MSP. Therefore, dual pricing system can be sustained only through non-price interventions such as infrastructural support for regional dispersal of growth in food grains output and adequate machinery for public procurement.

Apart from flexibility for quick decisions, the new system requires better market intelligence, e.g., forecasts as well as concurrent analyses on prices and sales in markets in India and abroad. There is thus an immediate need to put in place a suitable institutional mechanism for this purpose.

Times of India, June 30, 2006.

17

Organised Retailing

Impact on India's Farm Economy

DRIVEN by rising demand, Indian agriculture is getting increasingly diversified into high-value products like dairying, vegetables, fruits, poultry, meat, fish, etc. These products being input-intensive and perishable require a marketing infrastructure that enables the farmers to minimise post-harvest losses which are substantial—over one-third of output—and realise a much better share in the price paid by the consumers, which is now below 15 per cent of some horticultural products.

Wider adoption of even known technology by the farmers as well as incentives for further innovations crucially depend on a favourable marketing environment.

The recent decision of the Government of India on permitting foreign direct investment (FDI) up to 51 per cent in the multi-brand retail holds promise for speedy modernisation of organised retail in the country through improved technology and competitiveness.

Farmers, apart from other non-farm enterprises and consumers in general, would benefit significantly from the development of organised retail, especially in view of the policy requirement that at least 50 per cent of total FDI brought in should be invested in backend infrastructure. Much of this investment could go into agriculture produce market infrastructure, storage, ware-housing , cold chains, processing and so on.

Since the growth of food processing industry is demand-driven, experience the world over shows that its growth is highly correlated with the development of modern retailing. The slow growth of food-processing industry and slow growth of rural non-farm sector in the country are responsible for insufficient diversification of agriculture and excessive dependence of labour force on it.

Expansion of organised retailing requires supporting infrastructure such as storage facilities, assured electricity supply and transport and communication networks that can be provided mainly through public investment or through public-private partnerships. These are essential even for the policy of attracting FDI into retail to succeed.

The fears that the growth of organised retail would result in large-scale displacement of *kirana* stores, small traders and commission agents and pushcart vendors are highly exaggerated. Experience of developing economies in East Asia, including China, shows that even though the proportion of those employed in traditional retail would decline, their absolute number would continue to increase for two to three decades, if the economy grows at a minimum of 6-7 per cent per annum.

In the mean time, the growth of agro-processing, induced by modern retailing, and employment opportunities created by economic growth in general could absorb many of those displaced in the traditional sector. The resilience of the traditional retailers, including pushcart vendors, should not be underestimated. Their coexistence in large numbers with modern retail in several countries testifies to their ability to compete with modern retail by upgrading themselves and by becoming more consumer friendly.

However, policies for easy access to credit, training in skills and modernisation of their enterprises need to be put in place for enabling them to adjust to changes by upgrading their present activities wherever possible or move into emerging occupations like agro-processing. They also need to be freed from harassment by petty officials and police.

The tasks on the farming front are going to be far more challenging. Over 80 per cent of farmers now are small and marginal with increasing participation of women. Their awareness of the emerging marketing problems as well as their bargaining power while negotiating with more powerful buyers like organised wholesalers and modern retailers need to be raised by organising them into sales cooperatives and, where necessary, into farming cooperatives for ensuring supply of quality products in demand.

Organised retailers may prefer to provide credit and extension services to large and middle farmers on account of lower transaction costs, as they are more knowledgeable and resourceful and, though smaller in number, account for nearly half the total cultivated area. Easy access to institutional credit and publicly provided extension services to small and marginal farmers or cooperatives can improve their bargaining power with organised retailers. The East Asian experience shows that in the absence of strong public support systems, small farmers could be marginalised further, leading to sharp rural inequalities, unlike in the Green Revolution days.

Public policy can promote greater competition among the organised retailers or wholesalers by facilitating the entry of new players. The laws on contract farming need strengthening to improve the competitive position of farmers in relation to agro-processing firms and retailers providing quality inputs and extension services.

The existing Agricultural Produce Marketing Committee Acts, in vogue in the states, need to be considerably modified to allow farmers to sell their produce directly to the retailers and allow easy movement of marketable surplus across the states.

Economic Times, November 13, 2012.

18

Strategy for Soil and Moisture Conservation through Watershed Development

Introduction

Water resources management has been a major area of interest for Professor A. Vaidyanathan. The aspects on which he worked are wide-ranging—from major and medium irrigation projects to tank irrigation and pricing of irrigation water (Vaidyanathan, 1999; Government of India, 1992). Way back in 1991, he wrote a pioneering paper on soil and moisture conservation through watershed development (Vaidyanathan, 1991). Because the issues raised in this paper are of considerable relevance to the present concerns for reshaping the strategy of watershed development, I have chosen to write on this subject as a token of my tribute to him on the occasion of his 75th birth anniversary.

Watershed development has been conceived basically as a strategy for protecting the livelihoods of the people inhabiting the fragile ecosystems experiencing soil erosion and moisture stress. The aim has been to ensure the availability of drinking water, fuel-wood and fodder and to raise the income and employment for farmers and the landless labourers through improvements in agricultural production and productivity.

As an official programme, watershed development is over three decades old in India. It has undergone periodic reviews by expert committees with a view to improving upon its strategy and components. One such committee submitted its report to the Ministry of Rural Development in 1994 (Government of India, 1994). This technical committee had the benefit of studying and learning from the already existing cases of outstanding success in watershed development such as Sukhomajiri, Ralegaon Sidhi, Adgaon, certain tribal areas of Panchmahal, Mittemari and Jhabua. Learning from these and other success stories, the technical committee recommended planning on watershed basis through the participation of the people at all stages, active involvement of voluntary organisations [non-governmental (NGOs)], and coordinated effort resulting in convergence of treatment by the concerned government departments.

The experience from the ongoing watershed development programme for drought proofing of rain-fed agriculture by conserving land and water resources was encouraging in the initial period of a few years following the adoption of new guidelines in 1995 (Rao, 2000). The single most important factor accounting for the positive impact of this programme is community participation and decentralisation of programme administration made possible under these guidelines. This has come about on account of political and bureaucratic commitment to the new strategy. This encouraging experience points to the large potential for extending such participatory programmes under the government sponsorship.

Yet, the present level of participation seems highly inadequate for sustaining this development, especially in areas where the programme has proceeded too fast by fulfilling the 'targets' for completion of works without the required social mobilisation, institution building and leadership formation at the grass-roots level (Yugandhar *et al.*, 1999). Thus, post-project sustainability is very low in government projects when compared to those run in collaboration with the NGOs or by the NGOs on their own. The experience, on the whole, shows that the 1995 guidelines have not been given a fair trial, and we have still to go a long way to see the replication and translation of the individual success stories into an institutional success on a wider scale.

Recognising such gaps, the Ministry of Rural Development, Government of India, appointed in 2005 a technical committee on watershed programmes, under the chairmanship of S. Parthasarathy, to revisit the entire issue. The Committee's recommendations (Government of India, 2006) are being followed up by the Government by constituting the National Rainfed Area Authority (NRAA).

Scenario for 2020

Looking at the prospects for the year 2020 with the twin objectives of achieving the required growth in agricultural output and ensuring water security for drinking and other domestic purposes, it is clear from a number of studies done on the subject that the availability of groundwater would emerge as a critical bottleneck: the requirement of groundwater would significantly exceed its availability. Since the requirement will exceed the replenishable supplies for irrigation at the existing rate of recharge of groundwater, it may lead to over-extraction of groundwater in several states by 2020, if measures for recharging groundwater through rainwater harvesting are not successful (Chopra and Goldar, 2000).

Closely related to this is the problem of degradation of land in the rainfed or dry land areas. As it is, the rate of degradation of land in such areas in the 1990s is likely to have proceeded at more than twice the rate observed in the 1980s, basically on account of soil erosion from run-off (Reddy, 2000). This is happening despite all the official and NGO interventions for soil and moisture conservation through watershed development which may not have covered more than 25 per cent of rainfed area needing treatment. The acute famine for drinking water experienced in several parts of the country during the recent years shows that droughts are becoming increasingly severe. Clearly, the existing interventions for drought proofing have failed to neutralise the overall processes of degradation of natural resources underway in the rural areas. This calls for a thorough reappraisal of our agricultural policies with a view to evolving an overall policy framework that provides adequate incentives and opportunities for soil and moisture conservation through watershed development as well as for judicious use of scarce water resources in rainfed areas.

According to the Working Group on Watershed Development, Rainfed Farming and Natural Resource Management for the Tenth Plan, 107 million ha of land in the country are subject to degradation. A total of 27.5 million ha was expected to be treated under different programmes by the end of the Ninth Plan. The targets for Tenth and Eleventh Plans were 15 million ha and 20 million ha respectively. It is estimated that 88.5 million ha, including the forest areas needing afforestation, would have to be treated under watershed programmes by 2022, that is, by the end of the Thirteenth Plan, at a total cost of ₹ 72,750 crore. Such an upscaling of watershed programmes is essential for the restoration of the carrying capacity of degraded lands and to prevent further land degradation.

Impact of Watershed Development Programmes

A decade of implementation since the mid-1990s under the new guidelines is a long enough period to justify the evaluation of a nation-wide project like this. Unfortunately, however, no systematic effort has been initiated at the national level so far for undertaking evaluations either of a concurrent nature or in-depth studies by independent institutions adequately equipped for the purpose. What we have are some studies undertaken by certain NGOs and individual scholars from certain institutions on their own. Besides, we have a good deal of anecdotal information gathered by the senior officials of the concerned departments from the state and the central levels, which is authentic and useful. Much of this can be scanned from the tour notes of such officials; from the deliberations at high-level meetings and seminars; and from the published papers by individuals. The inferences in this paper are based on some such studies and anecdotal information, and on a series of useful discussions with knowledgeable persons on the subject.

A clear impression one gets from all this information is that the overall impact of the programme has been positive when compared to the period before the implementation of new guidelines. Crop yields have risen and there has been an increase in area under cultivation in the rabi season, leading to a rise in employment and reduction in migration of labour. Availability of fodder for animals has also improved leading to a rise in the yield of milk (Rao, 2000). Although there are reports of an improvement in the access to drinking water during drought in some project areas, the overall impression is one of acute distress in respect of access to drinking water.

Lessons from the Experience

Despite some improvement in the performance mentioned above, the experience so far clearly brings out the failure of the programme to make a desired impact. The experience from the on-going programme in general and from the few successful cases in particular raises a number of important issues which have a bearing on the effectiveness of watershed development programme and its sustainability when the present programme comes to an end.

The following major issues are relevant in this context: (1) social mobilisation, institution-building and leadership formation for ensuring effective participation of people on a sustained basis; (2) building equity into the programme by addressing adequately the livelihood concerns of the landless, the resource-poor farmers and women in general; (3) convergence of various disciplines, particularly agriculture development programmes, with watershed development; (4) strategy for conservation of water harvested as an integral part of watershed development; (5) capacity building through training at various levels; (6) expert and independent evaluation of the programme; (7) ensuring accountability of the programme administration to the stakeholders at various levels; and (8) integration of watershed development programme with wage employment programmes. These are discussed below in some detail.

Social Mobilisation, Institution-Building and Leadership Formation

Ironically, it appears that the very same factor, *viz.*, people's participation and decentralisation of programme administration, which accounts for whatever positive results achieved so far, is highly inadequate for sustaining this movement in many areas. This turns out to be the most glaring deficiency in the implementation of the programme so far. This is not surprising given the entrenched bureaucratic ethos and the nature of our social structure. Even with the best of intentions, it requires quite some time for the bureaucracy, especially at the lower levels, to undergo the necessary motivational and attitudinal transformation for plunging wholeheartedly into the task of empowering people by delegating authority. Again, owing to the mutually conflicting interests of different groups constituting our rural society, including issues such as unrecorded encroachments or faulty recording of areas under different crops, building up institutions and evolving leadership for rallying the rural community around common interests with adequate stake for weaker sections and sustaining the momentum is a painstaking and time-consuming task.

The neglect of social mobilisation is reflected in a number of deviations from the original guidelines in the implementation of the programme. For example, instead of spending the first one and a half years or one year in social mobilisation, as originally envisaged, hardly a few months and, in a large number of cases, only few weeks are devoted for this activity. Besides, the outlays on training and capacity building have been slashed under the revised guidelines. NGOs in general are better equipped to undertake the work towards creation of awareness, social mobilisation and capacity building, besides providing greater flexibility than government agencies. But the revised guidelines for Haryali introduced in April 2003, instead of evolving mechanisms at the national as well as regional levels for selecting the bonafide and competent NGOs, have severely restricted the role of NGOs as programme implementing agencies (PIAs) in watershed development. This has been done despite the mounting evidence that the performance of watersheds in the implementation of which government or externally aided agencies collaborated with NGOs has been distinctly better than that which have been implemented by the government agencies alone.

Further, the role of the general body of stakeholders represented by the watershed associations has been considerably curtailed with a much greater role assigned to the watershed committees. Therefore, the programmme performance in respect of social mobilisation needs a thorough review for initiating remedial measures with highest priority. In particular, measures would be needed for generating interest at the local level, such as participation of women's thrift groups and common interest groups including those for usufruct sharing from common lands. Also, in view of regional variations in resource endowments and the needs of the people, adequate flexibility has to be built into the guidelines by keeping open a substantial portion of the outlays for supporting activities emanating from innovative local effort.

Integrating the Concerns for Equity

Experience shows that equity in the distribution of benefits is a pre-requite for effective or broad-based participation and sustainability of the project. But the programme has failed to build adequately the livelihood concerns of the landless, the marginal farmers and women. Since watershed projects are land- and water-related interventions, those with access to land and water have a greater incentive and stake in the project. Landless labourers, on the other hand, have the least incentive except wages paid on project works, unless they are provided water rights. In general, the focus has been on water harvesting through check dams and other such structures utilised mostly by the resourceful farmers for growing water-intensive crops which have driven down the water table. This has led to the shortage of drinking water and hardships for women who now have to walk much longer distances to fetch water or draw water from deep wells where water levels have been driven down.

Contrary to the original guidelines, the ridges or the upper reaches in the watersheds, where the weaker sections of farmers tend to be concentrated, have been neglected. Similarly, the village common lands, which account for a significant part of the income derived by the poor, are rarely treated for regeneration. Livestock development and its fodder needs for the poor, water bodies for fishing and trees for the landless have also not received adequate attention. This experience suggests the need for appropriate revisions in the guidelines for pre-empting certain portion of outlays towards activities that meet the livelihood concerns of the poor as well as for ensuring equitable sharing of the benefits from watershed development such as water, fodder and usufructs from common property resources. This would also require adequate representation for the landless and other poorer households in the Water Users' Associations. The Parthasarathy Committee in its recommendations accorded a high priority to livelihood concerns by making explicit provision for such activities (Shah, 2006).

Converge of Various Disciplines including Agriculture Development Programmes

The integration of watershed development with Joint Forest Management was attempted following the decision by the Ministry of Rural Areas and Employment with the concurrence of the Ministry of Environment and forests: Forest lands forming part of such watersheds were also to be treated simultaneously as per the guidelines. Several state governments had taken the necessary steps in this direction. However, at the moment, there is little presence of agriculture department in the Drought Prone Areas Programme (DPAP) watersheds by way of promoting locally relevant research and extending suitable technologies, inputs and other necessary support to farmers. The Ministry of Agriculture has adopted the participatory approach underlying the DPAP guidelines in respect of the National Watershed Development Project for Rainfed Areas (NWDPRA) watersheds under its purview. This is a welcome move, as it would raise the productivity from Agriculture Department's own measures for raising agricultural production already in place in these areas. The convergence of services from Agriculture Department in the watersheds managed by the Ministry of Rural Development would contribute to raising the productivity of land and water resources augmented through water harvesting and would thus induce the farmers to sustain their effort towards watershed development in these areas.

However, the problem of lack of convergence of different disciplines is common to the watersheds, in general, because of the virtual withdrawal of departments not directly managing them. This has rendered the delivery of technical knowledge at the watersheds ineffective. An early integration of the interventions at the watershed level by different departments dealing with land, water, agricultural and other rural-based activities by bringing them together under one coordinating mechanism is very much called for.

Strategy for Water Conservation

The existing guidelines for watershed development are designed basically to augment water resources through water harvesting with virtually no stipulation for conserving the harvested water. On the contrary, the existing cropping pattern and the system of pricing of electricity for pumping water encourage overdrawal of water and its wasteful use. This is a major lacuna in the existing guidelines. This needs to be rectified by building necessary safeguards against wasteful use of water.

Since water-use-efficiency for irrigation is very low in the country, achievement of water security depends not only on the exploitation of the remaining potential, but increasingly on the efficient and productive use of available irrigation water (Government of India, 1999). There exists a vast gap now between the knowledge available for improving water-use-efficiency and its actual application in the country [Consultative Group on International Agricultural Research (CGIAR) 2001]. This is explained by the lack of incentives to follow improved practices under the existing system of management and pricing of water.

Awareness on the part of the farmers about the impending water crisis and their cooperation in pricing the water on a volumetric basis is essential for making the change-over politically feasible and sustainable. A possible solution, therefore, is to evolve a system where the village communities develop a stake in the conservation of resources. In certain situations, *panchayats* can regulate the cropping pattern and the use of water for raising water-use-efficiency and productivity. For example, they could prohibit the boring of new wells or impose taxes on the existing wells according to the paying capacity of farmers, or could impose a progressive tax on the cultivation of water-intensive crops. This may call for a change in the existing property rights and bringing ground water under the common property management regime (Reddy, 2006).

A subsidy at a flat rate, say, on the estimated consumption of electricity on holdings upto 1 or 2 ha, can be given to all the farmers or only to the small and marginal farmers, requiring them to pay for the extra consumption on a volumetric basis. This would have the merit of providing relief to a large number of deserving farmers in a manner that is politically feasible, even as there would be an incentive to conserve electricity and water. In addition, water-conserving cropping systems, technologies and practices need to be propagated in watersheds with necessary incentives, for example, through a subsidy on sprinkler irrigation.

Capacity Building through Training

Capacity building for watershed development through training encompasses wide-ranging tasks such as awareness building or imparting resource-literacy; development of technical skills; and reorienting motivations and attitudes of officials and political functionaries at all levels towards the need for empowering the people through decentralisation. Those who administer the programme, and who themselves need such training, are not best fitted to oversee these training programmes. While a number of measures have been taken for strengthening training at various levels since the adoption of new guidelines, experience so far strongly suggests the need for (a) broadening these programmes with a view to imparting training to bureaucracy at all levels as well as to political functionaries from *panchayat* level to parliament; (b) improving the content and quality of training programmes with due priority given to the processes of empowering people through decentralisation; (c) intensifying training by adequately taking care of the changing requirements in the field; and (d) ensuring the autonomy of training institutions and their uninterrupted functioning on the required scale by entrusting this task to an independent nodal agency at the national level and by guaranteeing adequate funding by the concerned ministries at the Centre.

Evaluation

Despite wide coverage of the programme in the country, there is virtually no institutional mechanism put in place at the national level for concurrent evaluation as well as for undertaking studies of basic and long-term nature. A programme like watershed development, which involves a hierarchy of administration and communities at the grass-roots level in highly varying agro-climatic and socioeconomic conditions, invariably requires mid-course corrections of varying degree in the light of ongoing experience. Typically, an implementing agency would see a greater value in spending an extra few crores of rupees for undertaking new works in the field rather than spending this money for research and evaluation. However, according to many observers, mid-course corrections can improve the programme benefits substantially, in some cases upto 100 per cent. But,even if we consider the improvement to be very modest, say, 10 per cent, even a 1 per cent of programme outlay on research and evaluation would have a very high pay-off in terms of achieving the programme objectives. It is of utmost importance, therefore, to put in place an institutional mechanism for research and evaluation in this field by involving reputed institutions in the country for upgrading the quality of evaluation, for example, by evolving methodologies for capturing the participatory processes and convergence of complementary interventions; by guaranteeing to them adequate funding on a long-term basis at the national level; and by ensuring for them autonomy in the pursuit of research and evaluation. In this context, the Parthasarathy Committee's recommendation for earmarking of 2 per cent of the overall programme budget and involving a wide network of socioeconomic research institutions for data collection, studies and evaluations is especially welcome (Vaidyanathan, 2006).

Such evaluations should include in-depth studies of successful watersheds with a view to drawing appropriate lessons for replication in other watersheds. Participatory evaluations by the stakeholders at the watershed level should be made an integral part of watershed development. It would be useful to entrust NGOs with monitoring and evaluation, apart from the task of project preparation.

Programme Administration and its Accountability to Stakeholders at Various Levels

There are, at present, multiplicity of watershed development programmes run by different ministries e.g., the Ministry of Rural Development, the Ministry of Agriculture, the Ministry of Environment & Forests and the Planning Commission. As it is, there have been some efforts to rationalise the guidelines for watershed development under different ministries. For example, the Ministry of Agriculture had adopted broadly the DPAP guidelines in respect of NWDPRA watersheds under its purview. It would be desirable for each department to review the performance of watershed development programme and revise the guidelines in the light of the experience gained from programmes under different ministries and the recommendations of the Parthasarathy Committee. A healthy competition between programmes run by different ministries should lead to the adoption of the best practices thrown up by experience.

At the watershed level, there is a need to ensure the accountability of management to the stakeholders. In many cases, records of money spent are not properly maintained. Proper accountability would require greater involvement of watershed associations, self-help groups and user groups in decision-making, e.g., planning, execution and financial management under the guidance of the PIA chosen. *Gram panchayats* should be concerned mainly with facilitating convergence, project review, monitoring and conflict resolution within the village community as well as across the neighbouring villages. A cluster approach in the selection of watersheds by different ministries would make for greater transparency as well as cooperation among the inhabitants of the contiguous watersheds and would strengthen the water users' associations federating at the block level. This would also make the provision of technical services, e.g. from agro-clinics, more economical.

Within the framework of the NRAA, there is a need at the district level for a steering or monitoring committee, possibly under the chairmanship of the zilla parishad chairman, and consisting of the representatives of different departments managing watersheds, reputed NGOs, social scientists from the local educational institutions and other public figures. Such an institutional mechanism would be helpful in taking a holistic view of watershed development with water conservation as a primary goal. Similar bodies could be constituted at the state level consisting of the representatives of the concerned ministries, NGOs, social scientists and public figures. These committees could periodically assess the performance of the programmes and take corrective measures.

Integration with Wage Employment Programmes

It would be desirable to integrate watershed development with the on-going wage employment programmes in operation in the area, particularly the National Rural Employment Guarantee (NREG) scheme. This would augment the resources for watershed development, even as it would contribute to the productive use of employment generated under the programmes. Such an integration would require co-ordination between various departments from the Centre to the district level.

It would be necessary for the Ministry of Rural Development and other concerned ministries to identify the degraded areas needing treatment under the watershed approach in the selected districts and launch a co-ordinated programme to integrate NREG scheme with watershed development.

Strategy for Rainfed Farming

Conceptually, the present strategy of watershed development is prompted by the need to protect the inhabitants of the fragile eco-systems from acute distress caused by recurring droughts. In this sense, it is a strategy for survival; even though in quite a few cases, the successful implementation of watershed development programme has led to a substantial increase in income and employment for the people. Even the many cases of outstanding success pioneered by the eminent NGOs reveal that such efforts were originally driven by acute distress caused by recurring droughts. Indeed, the overwhelming response from the people to the initiatives taken by the NGOs is explained by the misery to which they were subjected.

It is necessary to distinguish this strategy for protection or survival from a wider or holistic strategy for the development of agriculture in dry land or rainfed areas. Such a wider strategy would involve issues of agricultural technology suited to rainfed areas, price and credit policies, infrastructure development for processing, marketing and transportation of agricultural produce and trade policies to afford reasonable protection to the dry land farmers and to provide opportunities for export of their products. Some of these supporting measures are no doubt present in certain dry land areas, especially in the watershed areas run by some of the distinguished

NGOs, where there is a sustained increase in agricultural productivity and income. But this is nowhere comparable to the policy framework and infrastructure available to the areas in the country which have experienced the Green Revolution. Such a wider strategy for extending the Green Revolution to the rainfed or dry land areas is essential not only for deriving full benefit from soil and moisture conservation already accomplished in some areas through the watershed development programme, but also more fundamentally to provide the necessary economic incentives for turning watershed development into a truly spontaneous and demand-driven movement of the people in dry land areas not covered so far, and to sustain watershed development already undertaken when the current official programme comes to an end.

Social mobilisation, equitable distribution of benefits and convergence of various disciplines in watershed development, though necessary, would not be sufficient for optimising the benefits from soil and moisture conservation and for sustaining the interest of dry land farmers in such efforts in the long-run, unless there is a major shift in the existing policy for dry land agriculture. To appreciate this, it is necessary to understand the basic differences between the conditions faced by dry land farmers and those in the Green Revolution areas.

A large part of rainfed or unirrigated area is characterised by low and uncertain rainfall, low wages and high poverty. Owing to the unregulated exploitation of ground water, these fragile areas have been highly degraded. In the irrigated or favourable areas, given the technology, the decisions and efforts of farmers as individuals were central to the spread of the Green Revolution, whereas in the dry land areas, such individual farm-level decisions have to be preceded by community effort for soil and moisture conservation which is a painstaking and time-consuming process. Besides, dry land areas are characterised by greater agro-climatic diversity than the traditional irrigated areas, necessitating location-specific research for evolving viable technologies and practices. However, dry land areas have not received so far the priority they deserve in respect of agricultural research: the Green Revolution was characterised by land-saving but water-using technologies whereas dry land areas need water-saving enterprises and practices which optimise output per unit of scarce water. In the absence of such technologies, the farmers in the dry land areas go in for water-intensive crops like rice even when water resources are scarce, giving rise to conflicts on irrigation water and scarcity of drinking water. Unlike irrigated crops like wheat and rice, crops under dry land conditions face yield uncertainty as well as price uncertainty, in addition to being less remunerative, as price support and procurement operations are highly inadequate in these areas. Further, pricing of electricity for pumping water at a flat rate or supplying it free instead of on a volumetric rate has led to the over-exploitation of scarce groundwater resources. Infrastructure like roads and marketing is highly inadequate in dry land areas.

However, given the necessary shift in the policy framework in respect of agricultural research, extension and supply of inputs like seeds, infrastructure development and price and credit policies, dry land areas hold considerable promise for the development of agriculture. Indeed, the returns on investments in inputs as well as research at the margin are now higher for such areas than for irrigated areas, which so far have received greater attention (Fan and Hazell, 1997). Consumer demand is shifting fast towards water-saving enterprises like dairying, animal husbandry, horticulture and floriculture in which rainfed areas have a comparative advantage. Also, such products have a good potential for export. Agro-processing, therefore, holds a great promise in dry land areas. All of this combined can be expected to provide the necessary incentives for the conservation of natural resources and indeed for easing pressure on such resources. Watershed development can certainly become more manageable under such a scenario.

References

Chopra, Kanchan and Bishwanath Goldar (2000). "Sustainable Development Framework for India: The Case of Water Resources" (mimeo), Institute of Economic Growth, Delhi.

Consultative Group on International Agricultural Research (CGIAR), Technical Advisory Committee (2001). "Water and the CGIAR", Discussion Paper, CGIAR, Washington, DC.

Fan, S. and P. Hazell (1997). "Should India Invest More in Less-Favoured Areas?", International Food Policy Research Institute, Washington, DC.

Government of India(1992). *Report of the Committee on Pricing of Irrigation Water*. New Delhi:

Planning Commission (Chairman: A. Vaidyanathan).

———. (1994). *Report of the Technical Committee on Drought Prone Areas Programme and Desert Development Programme*. New Delhi: Ministry of Rural Development.

———. (1999). *Report of the National Commission for Integrated Water Resources Development Plan*, Vol. I, New Delhi: Ministry of Water Resources (Chairman: S.R. Hashim).

———. (2006). *From Hariyali to Neeranchal: Report of the Technical Committee on Watershed Programmes in India*. New Delhi: Department of Land Resources, Ministry of Rural Development (Chairman: S. Parthasarathy).

Rao, C.H.H. (2000). "Watershed Development in India: Recent Experience and Emerging Perspectives", *Economic and Political Weekly* XXXV (45), November 4.

Reddy, Ratna (2000). "Land Degradation in India: Extent, Costs, Determinants and Trends", (mimeo), Hyderabad: Centre for Economic and Social Studies.

———. (2006). "Getting the Implementation Right: Can the proposed Watershed Guidelines Help?", *Economic and Political Weekly* XLI (40), October 7-13.

Shah, Mihir (2006). "Overhauling Watershed Programme: Towards Reforms", *Economic and Political Weekly* XLI (27 & 28), July 8-14/15-21.

Vaidyanathan, A. (1991). "Integrated Watershed Development: Some Major Issues", Founders Day Lecture, Society for Promotion of Wasteland Development, New Delhi.

———. (1999). *Water Resource Management: Institutions and Irrigation Development in India*. New Delhi: Oxford University Press.

———. (2006). "Restructuring Watershed Development Programmes", *Economic and Political Weekly* XLI (27 & 28), July 8-14/15-21.

Yugandhar, B.N., J. Venkateswarlu and Vijay Kochar (1999). "Watershed Based Development in Arid and Semi-Arid Areas of Andhra Pradesh", *Journal of Rural Development* XVIII(3), NIRD, Hyderabad.

Published in Kadekodi, Gopal K. and Brinda Viswanathan (ed.) (2009). *Agricultural Development, Rural Institutions, and Economic Policy: Essays for A. Vaidyanathan*. New Delhi: Oxford University Press.

19

Sustainable Use of Water for Irrigation

WATER resources are becoming extremely scarce. Even after fully exploiting the usable water resources by the middle of this century, the supply-demand balance for irrigation water can be achieved only by improving irrigation efficiency in a big way. Interestingly, water savings in crop production the world over have so far accrued indirectly from the rise in crop yields and very little directly from improvement in water-use-efficiency. Focus on productivity per unit of water used is new for the international agricultural research system. Nearly 200 years ago, Thomas Malthus focused on the scarcity of land in relation to population growth as the cause for recurring famines. This perception prompted the evolution of land-saving technologies which substantially raised productivity of land. Until recently, irrigation water was hardly regarded as a scarce resource.

Despite the Green Revolution, food grains output in the country grew at about the same rate as gross irrigated area in the 1970s, so that there was no growth in productivity of irrigation water for food grains. During the 1980s, however, the Green Revolution was broad-based covering rice in the central and eastern regions of the country where high rainfall supplements irrigation. Thus food grains output grew at a higher rate than irrigated area resulting in a rise in the productivity of irrigation water. However, in the decade of the 1990s, there was a deceleration in the growth of food grains output leading to a significant decline in the productivity growth of irrigation.

What are the major sources for raising the productivity of irrigation water? Genetic improvements which raise the productivity of land can, at the same time, raise water productivity, especially if such genetic improvements are targeted towards saving water.

These improvements include the development of crop varieties with better tolerance to drought, cool seasons which reduce evaporation and evapotranspiration, and saline conditions. Biotechnology has a great potential for raising yields and imparting stability in adverse environments.

Secondly, productivity of irrigation water can be raised by skillfully supplementing it with rain water, especially in the high rainfall areas. Agricultural development strategy that accords greater priority to high rainfall areas like the central and eastern regions in the country may result in significant savings on high-cost surface irrigation.

Third, water losses that occur from the point of delivery from the system to the farmer's field due to evaporation, flow of usable water to sinks, pollution, salinity and water-logging can be minimised with appropriate management practices, provided there are adequate incentives for adopting water-saving practices. Fourth, as incomes rise, consumer demand shifts away from some of the water-intensive crops like rice to water-saving horticultural products. However, trade and price policies and policies on input subsidies, including on irrigation water, would determine whether farmers would switch over to water-conserving enterprises. Finally, globalisation opens up opportunities for countries faced with water scarcities to specialise in the production and export of water-saving crops and import water-intensive ones.

Irrigation Management Transfer to the user-farmers is being increasingly practiced the world over to provide correctives to the distortions arising from the failure of the market as well as the state. India has initiated Participatory Irrigation Management (PIM) through the Water Users' Associations. Conservation of groundwater is another area where market failure as well as government failure can be glaring. As a solution to this, Participatory Watershed Development is being practiced in many countries, including India, for the sustainable use of land and water resources.

In India, the performance of PIM so far indicates that the unwillingness to fully transfer powers and functions to the user associations is the major bottleneck. Similarly, the experience with watershed development under the participatory approach in India shows that the immediate results are no doubt positive in terms of the rise in water table, crop yields, employment and incomes of the participants. But there is considerable uncertainty regarding the sustainability of these institutions, that is, their ability to hold different social groups together and mobilise resources from the community for maintenance of structures already created and for further development when the assistance under the officially sponsored programmes ceases.

Price policy of the government, by influencing cropping pattern, has an impact on the productivity of water. A glaring illustration of this is the price policy for rice and wheat which has resulted in the accumulation of over 60 million tonnes of food grains with the Food Corporation of India (FCI). For each hectare of paddy raised under irrigated conditions, as much as 16,000 tonnes of water is applied in the semi-arid North-Western Indian states, from where bulk of the rice is procured by the FCI. Water-saving technologies are not adopted by the farmers because water is available at a very low cost or is even free. For example, despite considerable potential for saving water through drip and sprinkler irrigation, only about 1 per cent of irrigated area in the country is covered by these methods.

The debate on water rates in the country so far has centered round the revenue accruing to the government for covering the operation and maintenance charges. The objective of conserving water needs to be brought into the forefront in these discussions.Water Users' Associations and *panchayats* should be fully empowered and induced to charge and collect water and electricity rates from the farmers on the basis of the volume of actual consumption where metering is possible or on the basis of the quantities of water received or electricity consumed as estimated by the farmers' associations where metering is uneconomical or otherwise infeasible. The revenues so collected should be allowed to be retained entirely by the Water Users' Associations under the aegis of *panchayats* for maintenance and development of irrigation services. Given the technology and public policy, institutions concerning water use hold the key to raising water productivity by bridging the vast gap that now exists between knowledge and its application.

Keynote Address at the Final Review Workshop organised by the Indira Gandhi Institute of Development Research in Association with University of Hyderabad, Hyderabad, August 20, 2002.

III

Andhra Pradesh

20

Andhra Pradesh

Fifty Years of Development and Challenges Ahead

ANDHRA PRADESH (A.P.) has made considerable strides in development in the last 50 years. This is in conformity with the national trends in development in the post-independence period. Per capita income in the state has more than tripled over this period and social development indicators like literacy level and life expectancy improved significantly with a substantial decline in poverty.

But, does A.P. stand out prominently in respect of its achievements when compared to the national average? Or has A.P. adopted any of the best practices making impact in other states? Unfortunately, the answers to such questions are not in affirmative. In fact, our growth performance has been just around the national average until recently. There are even signs of sliding down in our performance. For example, between the decades of 1980s and 1990s, among the 14 major (non-special category) states, A.P.'s rank in respect of growth in per capita gross state domestic product (GSDP) slided down from 5 to 8, whereas all the other three southern states have either maintained or improved their rank: Karnataka moved up from no. 7 to 1, Kerala from no. 13 to 7, while Tamil Nadu maintained its high rank at 3. During the same period, West Bengal scaled up dramatically from 11 to 2.

The position has been even worse in respect of human development index comprising indicators like literacy rate, life expectancy at birth etc. A.P.'s rank in this respect slided down from 9 to 10, while Kerala, Tamil Nadu, Karnataka and West Bengal maintained their respective ranks at 1, 3, 7 and 8. The state also has a dubious distinction of being at the top in regard to the incidence of child labour and the number of suicide deaths of farmers. A.P.'s slide down in respect of per capita

GSDP as well as social development is explained by its low level of social and economic infrastructure: its rank in this respect in 1995 was 10 as against 2 for Kerala, 4 for Tamil Nadu, 9 for Karnataka and 5 for West Bengal.

Capital formation in agriculture and irrigation in the state suffered badly over the last 25 years, especially in the backward Telangana and Rayalaseema regions. Public investment in health and education has also not been adequate. No doubt, the state has been facing resource crunch, but this is partly because it has been incurring unduly large expenditure on subsidies on rice, water and power, whereas targeting these subsidies strictly to the deserving sections like small and marginal farmers or to those below the poverty line could have released substantial amounts for capital and social expenditures.

A.P. also failed to release the creative energies of people at large by triggering-off social transformation in rural areas like in Kerala, West Bengal and Karnataka, by its neglect of land reforms in areas with feudal antecedents like Telangana and its extreme reluctance, unlike in the other three states, to decentralise rural development administration through the Panchayati Raj Institutions.

Because of these basic structural deficiencies, the economic reforms initiated in the state in the late 1990s or the boost given to the information technology (IT) sector, though desirable, could not have suddenly secured for the state an enduring place in the group of 'progressive' or 'performing' states. However, a silver lining in the state's performance is the emergence of women power exemplified by the self-help groups for securing access to credit and for organising income earning enterprises such as by the Development of Women and Children in Rural Areas (DWCRA) groups. The rise of women power is also reflected in a significant decline in the fertility rate over the 1990s. Moreover, whenever opportunities were provided, the poor in the state have displayed a commendable capacity for successful social action, as in the case of rainwater harvesting through Watershed Development, water conservation through Water Users' Associations and protection of environment through Joint Forest Management. If some of these ventures eventually met with failure, it was more due to bureaucratic meddling and intrusion of politics into their functioning.

The way ahead for A.P. is not going to be easy. The present government has rightly accorded a high priority for irrigation, agriculture and rural development. But land reforms, however modest, expansion of employment opportunities in the rural non-farm sector and ensuring regional equity in access to irrigation and jobs are the major challenges facing the state.

About 70 per cent of the existing irrigation in the Telangana and Rayalaseema regions is served by wells. This is now sustained by free power. Beyond a point, this policy would be counter-productive as the water table depletes further and also results in resource crunch. Giving highest priority for the full exploitation of the remaining potential from major and medium irrigation projects in these two regions with proper rehabilitation of displaced persons is the only lasting solution. A.P. also needs to decentralise development administration by fully devolving functions, finances and functionaries to the elected local institutions.

The New Indian Express, June 30, 2006.

21

Andhra Pradesh

Fastest Improver in Economic Freedom and Growth?

A HIGH correlation between the expansion of economic freedom and accelerated growth in gross domestic product (GDP) is well established in India since the early 1990s when economic reforms were introduced. But there are inter-state variations in this correlation depending on the differences in the initial levels of infrastructure development and the governance patterns. A recent publication "Economic Freedom of the States in India 2011" partnered by the Cato Institute, Friedrich Naumann Foundation, Indicus Analytics and Academic Foundation throws new light on such relationships by using 20 indicators classified broadly under three parameters, *viz.*, size of the government, legal structure and protection of property rights and regulation of credit, labour and business, for ranking the states.

Swaminathan Aiyar, in his essay, rates Andhra Pradesh as the fastest improver in economic freedom, faring exceptionally well in regulation of labour and business, between 2004 and 2009 when Dr Y.S. Rajasekhara Reddy was the Chief Minister. Improved business climate resulted in state accelerating its gross state domestic product (GSDP) growth to higher than the national average. Aiyar also demonstrates convincingly, from the Andhra Pradesh experience, the compatibility of high growth with programmes for poverty alleviation and social justice, primarily on account of generation of revenues from high growth and eventual decline in subsidies because of the decline in the share of agriculture in GSDP and decline in poverty. Thus growth is a necessary condition for achieving inclusiveness, and inclusiveness in turn becomes necessary for sustaining growth politically, as demonstrated by Dr Reddy by returning to power in 2009—a lesson that the Congress Party at the national level had learnt the hard way, 15 years after piloting reforms when at the time of the 2004 elections it promised 'economic reforms with human face'.

Impressive as the performance of Andhra Pradesh is, it needs to be understood in a wider perspective. First, Andhra Pradesh had already been on the reforms mode since the mid- 1990s under the leadership of Shri Chandrababu Naidu. Several reforms were initiated affecting the information technology (IT) sector, power sector, management of irrigation, conservation of water through watershed development and state finances. As a result, the GSDP growth rate of the state, which remained below the national average for over four decades of planned development, pierced the national average for the first time during 2001-02 to 2004-05, when the GSDP growth rate of the state rose to 6.32 per cent against the all-India growth rate of 5.99 per cent. The per capita GSDP in the state also crossed the national average for the first time in this period. The lagged effects of reforms introduced earlier helped in Dr Reddy's period, apart from his own dynamism and reforms.

Second, high agricultural growth—twice the national average—achieved in Dr Reddy's period cannot be traced to economic freedom index, as there is no indicator bearing on economic freedom in this sector but, as Aiyar says, did contribute significantly to accelerating GSDP growth. By neglecting agriculture, Sri Naidu left considerable slack in this sector, and by restoring priority to agriculture through the sheer revival of the traditional public support systems like extension and better access to credit and inputs, Dr Reddy allowed this latent potential to express itself. Similarly, control of Maoism is also not traceable to the index of economic freedom as constructed, but, as demonstrated by Aiyar, was responsible for high growth in the northern districts of the state. Expansion of irrigation in this period cannot be traced to the massive investments made which have yet to yield results, but to generous subsidies for well irrigation in the form of free power which, as Aiyar rightly observes, have led to unsustainable exploitation of ground water. For example, during this period nearly two-thirds of increase in irrigated area came from wells. These areas are known for high value activities like dairying and horticulture which contributed significantly to high agricultural growth.

Third, a disquieting feature of the present index of economic freedom is the absence of indicators bearing on inclusiveness in development such as expansion of freedoms and opportunities through expenditures on education and health, and accountability in governance in rural areas by empowering people. The present index is in fact biased against human development, as it clubs together all the revenue expenditure, any increase in its ratio to GSDP signifying lower economic freedom. For Andhra Pradesh in the last two years, revenue expenditure on health and education constituted 97 per cent to 98 per cent of total expenditure on these heads. This accounted for nearly one-fourth of total revenue expenditure. The Planning Commission at the time of the formulation of the 11th Five Year Plan had expressed its reservations on the drive for bringing down the revenue deficit to zero by the year 2009, as it could hit the social sectors like health and education.

While Andhra Pradesh succeeded in accelerating its GSDP growth rate to higher than the national average, during the same period, its performance in respect of human development was sliding down below the national average, as brought out by the Centre for Economic and Social Studies (CESS), Hyderabad, in its recent work. Apart from lower public expenditures, lack of accountability to local communities is the main factor. Engaging contract teachers for schools in the state was a right move, as Aiyar points out, but improvement in the system as a whole is possible only by ensuring accountability. A recent study by the National Council of Applied Economic Research (NCAER) on the devolution of powers to Panchayati Raj Institutions, based on the information from the state governments, places Andhra Pradesh just around the national average, but going by the field data, it ranks below the national average and certainly below the three southern states.

Finally, the post-reform experience brought out by Aiyar and others in this report shows considerable fluctuations in the ranking of states in respect of economic freedom, those ranking high in one period sliding down to low level in the subsequent period, suggesting that some of the reforms tend to disappear with the exit of imaginative leaders. Therefore, institutionalisation of reforms and their insulation from the political exigencies of the day is a matter for serious reflection.

Based on the presentation made by the author at the Conference on "Fastest Growth in Andhra Pradesh–Economic Freedom Index 2011", organised by the Federation of Andhra Pradesh Chambers of Commerce and Industry and Friedrich—Naumann Stiffung fur die Freiheit, at Hyderabad, 29 July, 2011. Published in *Economic Times*, September 9, 2011.

22

Agribusiness Potential of Andhra Pradesh

BECAUSE of its favourable agro-climatic conditions and central location, Andhra Pradesh offers considerable potential for the growth of agribusiness. But this growth can be expected only on the solid foundation of food grain security and rising farm productivity. For example, the deceleration of agriculture growth during the decade 1995-2004 was associated with slowdown in agro-processing and exports of agro-products.

Marketable surplus of horticulture produce holds the key to the growth of agribusiness. Experience shows that more than relative prices, yield levels of horticulture produce determine the proportion of output marketed. But yields of horticulture produce in Andhra Pradesh are only about half of yields in other progressive states in the country. This points to the significant potential for raising farm yields. Besides, only about 2 per cent of fruits and vegetables produced are processed in the country as against over 60-70 per cent even in developing economies of Asia and Latin America. The demand for the processed products is going to be high as per capita incomes rise and urbanisation increases. Thus the growing size of the domestic market, by itself, can be the major determinant of the growth of agribusiness.

The constraints are mainly on the supply side ranging from low yields to inadequate rural infrastructure like roads, power, storage and marketing. Added to this, seasonality of production, uncertainty of yields and hence unpredictability of raw material supply result in low margins in agro-processing and agribusiness in general. Further, the preference of consumers for fresh fruits and vegetables presents a rare case of the finished or processed product having to compete with its own raw material. This is because, unlike in the developed countries, the costs of processed foods are much higher than fresh foods in India owing to various supply bottlenecks, technology, scale and taxes.

Taxes on horticulture products are among the highest in India, while they are nil in the developed countries. Even packaging material is taxed in India accounting for over 20 per cent of total costs. State-level taxes like VAT, central tax, entry tax, octroi, mandi tax, etc., further add to the costs. Thus consumer preference for fresh rather than processed foods in India is easily explained by their high relative prices rather than factors like culture, tradition and tastes.

Agro-processing is characterised by lower capital- labour ratio with an employment elasticity at least two and a half times higher than for manufacturing in general. Hence employment potential in agribusiness is quite high with a large scope for employment of women.

Export of agro-products is no doubt important in the initial stages but saturation of markets may soon be reached in the developed countries. On the other hand, technology and capital may be flowing from developed economies into the economies like India because of expanding domestic market. But so far foreign direct investment (FDI) has been flowing not so much into core areas as into retailing, brewing liquor, mineral water, etc.

To exploit the full potential from this industry, certain important interventions are needed in respect of technology, infrastructure and institutions. Biotechnology has made considerable strides to suit the needs of quality in processing, storage and preservation.

This needs to be tapped. The infrastructure for cleaned and graded produce has to be strengthened as consumers become increasingly health conscious and show preference for graded agro-products. Training in appropriate skills is necessary to meet the growing needs.

Promotion of contract farming requires appropriate legislation to enforce contracts. Also, Agricultural Marketing Acts need to be amended to suit the changing needs. Provision of credit is recognised as a priority area but the share of agribusiness in institutional credit is only about one-third of its share in industrial output. This needs to be redressed. Insurance is another area which has to be addressed for improving the environment for investment in this sector.

Based on the Keynote presentation made at the Seminar on Agribusiness Potential of Andhra Pradesh, organised by the Indian Society of Agricultural Marketing, at the Acharya N.G. Ranga Agricultural University, Rajendranagar, Hyderabad, 29th July, 2008. See, *Agribusiness Potential of Andhra Pradesh* (Seminar Proceedings), Indian Society of Agricultural Marketing, 2008.

23

Land Reforms in Andhra Pradesh

Major Issues

LAND to the landless—for cultivation as well as housing—which appeared high on the political agenda in Andhra Pradesh four years ago, seems to have taken a back seat, of late, and even forgotten in the official circles. The present government soon after coming to power in 2004 constituted the Land Committee under the chairmanship of Shri Koneru Ranga Rao, a Minister, "to assess the overall implementation of land distribution programmes of the Government and suggest measures for its more effective implementation". Out of the 104 recommendations made by the Committee, the Government had accepted 90. Quite a few of these will have far-reaching consequences if implemented effectively.

The current land movement in the state arose and gathered momentum basically on the demand for house sites for the landless poor in the urban and semi-urban areas. This was not a major issue before the Committee. Ironically, the rising demand for urban land is traceable, in a significant measure, to the massive programme for irrigation development called 'Jalayagnam', which is expected to ease pressure on land by raising agricultural productivity. Much before the water started flowing into the fields, apparently, a good part of the money injected into the system for irrigation projects experienced a reverse flow into the urban land market through the contractors and real estate dealers, raising the demand for land. Also, there has been a massive allotment of land in the outskirts of Hyderabad city and around other towns for the development projects.

The tremendous rise in demand for urban land led to the sky rocketing of its prices, making land out of reach for the urban poor and the middle classes. Many of the poor farmers in the outskirts of the city suddenly found themselves 'dispossessed' through the allurement of money which turned out to be a pittance when compared to the prices at which their lands were ultimately sold by the dealers. This generated a great sense of insecurity among the poor—those losing land as well as those in need of land for housing—in and around the urban areas.

Admittedly, Andhra Pradesh Government has been in the forefront in respect of building houses for the poor. But this cannot alleviate the misery of the poor if there is a policy failure to contain the prices of urban land and to pre-empt the land for the poor. Had the state government acted in time to acquire sizeable chunk of land for expanding the on-going programme of constructing houses for the poor as well as for direct allotment of house sites to them, the situation would not have turned so adverse.

The Left parties clearly saw the rapidly emerging scarcity of urban land and its adverse consequences for the poor. Already, their movement for house sites for the poor appears to be having some sobering impact on land prices by sending the right signals to the land market. The welfare impact from containing land prices would indeed be enormous for the poor and the less affluent in urban areas. Moreover, with growing urbanisation and migration of rural poor to the urban areas, the need for house sites for the poor will increase. While 'land to the tiller' is still a live issue, 'land to the dweller' is going to assume increasing importance, especially in urban areas.

The significance of the Land Committee Report should be seen in the light of the growing scarcity of land and the rising land values. The Committee's recommendations are, on the whole, quite moderate and not 'radical'. They basically centre round the more effective implementation of quite a few of the existing laws. If they sound 'radical' on occasions, especially on tribal land issues, it is because the Committee was constrained to recommend strict implementation of some of the existing laws dating back to 1917.

The three major concerns addressed by the Committee are: augmenting the 'surplus' land over the existing ceiling on land holdings as well as from government sources for distribution to the landless poor; encouraging investments by the tenants for raising productivity and incomes; and restoring to tribal population lands appropriated by the non-tribals.

A major recommendation of the Committee for the restoration of the 'assigned' land to the original assignees or, where it s not possible, its resumption by the Government for making fresh assignment to the landless poor has received wide support. However, the government while accepting this, rejected the related recommendation to include the *sarpanch* and president and secretary of the concerned Village Organisations of poor women in the Assignment Committee. Such an involvement is necessary for releasing a good chunk of land for the poor, on this account, without making the implementation process harsh and cumbersome, particularly because a piece of land would have changed a number of hands over the decades, the present occupants themselves being poor in many cases.

The Government has accepted the recommendations on ceiling surplus land and tenancy—the two major concerns in non-tribal areas. The Government has agreed to "reopen cases of Land Ceilings, wherever the cases have been decided basing on fraud and misrepresentation of facts". The number of such cases and the amount of 'surplus' land likely to be available are not known. But the indications are that, if effectively implemented, this measure could well unearth sizeable area. Of much greater potential, perhaps, is the acceptance of another recommendation by the Government to reassess the 'surplus' land consequent on the construction of irrigation projects, which resulted in a significant increase in irrigated area over the last few decades. It is highly unlikely, however, that the 'political will' needed to undertake such measures would be forthcoming.

In regard to the distribution of government as well as ceiling surplus land, the Government has not accepted the Committee's recommendation that "the maximum extent of land which may be allotted to a single individual be limited to 1 acre of wet land or 2 acres of dry land", as against the existing provision of 2.5 acres of wet land and 5 acres of dry land. This recommendation appears reasonable considering the significant rise in productivity and value of land as well as in the number of landless poor in need of land.

The Committee has noted that the provisions in the existing Tenancy Acts "have given rise to informal tenancy—almost 100 per cent of tenancy that exists is informal", and 55-60 per cent of the lands surveyed in the Delta region are under lease. Therefore, taking a realistic view, the Committee recommended that "a loan eligibility card should be issued to the tenants to enable them to access institutional loans so as to garner better gains from cultivation of lands, and the landlords on the other hand are not paranoid about losing their lands if the tenancy is recorded". Since the Government has accepted this recommendation, if implemented effectively, one can expect a rise in investment, productivity and incomes of the tenants.

Land alienation in tribal areas is remarkably high in Andhra Pradesh. According to the Committee, in tribal areas, "non-tribal population hold as much as 48 per cent of the lands. Every year more and more lands are passing into the hands of non-tribals and if it is not checked with a strong executive action, very soon the tribals may not have lands at all".

The Committee has made a series of recommendations for restoring lands to the tribals and for checking further alienation of their land. The Government has accepted the important recommendations, e.g. reopening and re-examining of orders in favour of non-tribals and review of a large number of cases of illegal occupation by non-tribals in some districts. However, a pre-requisite for effective implementation, according to the Committee, is the improvement in tribal land administration by putting in place field machinery with adequate staff, especially in Agency areas, and effective monitoring and review.

But, mere administrative capability is not sufficient unless the requisite political will is forthcoming. This is amply demonstrated by the massive failures in the implementation of Land Ceiling and Tenancy Acts in the non-tribal areas where administrative capability is not a bottleneck. The pulls and pressures from the vested interests can render even the most progressive land laws infructuous. This experience has prompted the Left parties to demand the constitution of an independent commission for implementing these recommendations.

But the Government has not agreed to this demand presumably in the belief that in a democratic polity with an elected government, a parallel authority on land issues may not be workable. Instead, it appointed a commissioner in the revenue department for implementing the recommendations. Clearly, however, there is a need for transparency and public accountability on this issue of vital importance.

The least that the Government could still do, to ensure effective implementation, is to constitute a high level committee, consisting of important stakeholders, to continuously monitor the implementation with powers to receive representations and bring up specific cases. This would be quite in line with the recommendation made by the Land Committee to institute such a mechanism for tribal areas—already accepted by the Government—in addition to the measures for awareness building among the affected people.

A major issue currently debated in the state is the massive sale of government land to the private parties for setting up industries and to the real estate dealers. The State Government has set up an ambitious target for mobilising revenues through this route—around 13 per cent of total revenues in 2008-09. This additional resource mobilisation target seems to explain a steep rise in the size of Annual Plan for the current year approved by the Planning Commission.

The Government needs to release the data on the extent of public land available and that already sold as well as proposed to be sold, location-wise. This is necessary for assessing whether the Government would be in a position to secure sufficient land for public purposes in future—for offices, social infrastructure like schools and hospitals and for meeting the housing needs of the poor in the context of growing urbanisation.

The Hindu, September 13, 2007.

IV

Telangana

24

Srikrishna Committee Report on Telangana

Overview

The Committee headed by Justice Srikrishna, constituted by the Government of India to examine the situation in Andhra Pradesh with reference to the demand for a separate state of Telangana as well as the demand for maintaining the present status of a united Andhra Pradesh, has submitted its Report after holding wide-ranging consultations with all the stakeholders and carrying out a comprehensive analysis of the relevant data on socioeconomic aspects. Among other things, the Committee examined the three basic issues bearing on the formation of a separate state of Telangana, *viz.*, its economic viability, social inclusion and cohesion, and wishes of the people of Telangana, and found them strongly favouring the formation of a separate Telangana state with Hyderabad as its capital. But the Committee has been deterred by fears on three counts: Vehement opposition from the influential sections of Seemandhra, particularly to Hyderabad becoming the capital of Telangana; similar demands for creation of smaller states cropping up in the country; and the impact on internal security situation with the anticipated growth of Naxalism and religious fundamentalism—discussed in Chapter 8 of the Report but withheld from public at large and submitted to the Ministry of Home Affairs in a separate cover along with the Report. The Committee tended to regard big states as strong and small as weak which could undermine national unity. It saw Naxalism essentially as a 'law and order' problem (which presumably can be 'faced effectively' in bigger states), rather than a socioeconomic problem calling for inclusive development which can be better accomplished in smaller states.

Balancing various considerations, the Committee recommended keeping the state united by creating a statutorily empowered Telangana

Regional Council, a Water Management Board and an Irrigation Project Development Corporation, as the most workable option under the circumstances (option 6 in the Report). As the second best option (option 5 in the Report), the Committee recommended the bifurcation of the state into Telangana and Seemandhra as per existing boundaries with Hyderabad as the capital of Telangana, and Seemandhra to have a new capital. This option is to be exercised only in case it is unavoidable and if the decision can be reached amicably amongst all the three regions by instilling confidence in the people of coastal Andhra, Rayalaseema and others who have settled in Hyderabad and other districts of Telangana with regard to the safety and security of their investments, properties, livelihood and employment (ch.9: 450-8).

The 'most workable' option recommended by the Committee basically amounts to restoring *status quo ante*, experimented but failed miserably as documented by the Committee itself. The Committee did not face the question squarely as to why this particular arrangement and several other assurances given to Telangana failed to work in the past and what prompts it to believe that a variant of such an arrangement would be workable in future. The Committee did seem to take note of the basic lesson from the experience that in a democracy like ours such 'safeguards' cannot withstand the pressures exerted by the 'interest groups' from the politically dominant region but, apparently on balance of considerations, chose not to allow this to influence its recommendations.

Surprisingly, in none of the two options did the Committee recommend safeguards for Rayalaseema—rightly regarded by it as the most backward region in the state. Further, whereas in the case of the 'first best' option, it elaborated the recommended safeguards for Telangana in great detail, in the case of the 'second best' option, it did not recommend any plan of action and a road map in regard to the assurances for the people from coastal Andhra and Rayalaseema settled in Telanganaa.

Growth and Inequality

At the time of the formation of Andhra Pradesh, average per capita gross state domestic product (GSDP) for Telangana was significantly lower than for coastal Andhra as well as for Rayalaseema. Starting with a low base, Telangana showed a 'catching-up process' resulting in a higher growth rate than the other regions but the level of its per capita GSDP continued to be lower than in the other two regions for at least four decades after the formation of the state. It is in the post-reform period that it overtook Rayalaseema, and now its per capita GSDP (excluding Hyderabad) stands above that for Rayalaseema and is next to coastal Andhra (Rao and Shastry, 2009). The analysis of growth by the Srikrishna Committee indicates a similar trend. The Committee does note, however, some continuing concerns in Telangana regarding public employment, education, and water and irrigation.

An important factor contributing to the catching-up process in Telangana is the in-migration of capital, enterprise and skills from Seemandhra—more so in the post-reform period—especially from coastal Andhra, into different districts of Telangana, prominent among them being Hyderabad and the adjoining areas. This is amply borne out by the data on inter-regional migration (Ibid.). Therefore, an analysis of how the benefits from rising gross domestic product (GDP) growth and employment have been shared between different social groups is essential to understand the rising discontent in Telangana, especially in the post-reform period. It is also necessary to understand the other qualitative aspects of growth such as costs of investment, uncertainty of returns and sustainability. For example, as pointed out by the Srikrishna Committee, agriculture in Telangana has become heavily dependent on well irrigation which is both costly and uncertain. This indeed accounts for the disproportionately large number of suicides by farmers in Telangana (Galab *et al.*, 2009)—also documented by the Srikrishna Committee.

The Committee notes the rising inequality in rural Telangana in the post-reform period [between 1993-94 and 2004-05, based on the sample surveys conducted by the National Council of Applied Economic Research (NCAER)] when the scheduled castes (SCs), scheduled tribes (STs), and minorities have suffered a decline in income: "Such deepening inequality in Telangana can not only sustain the separatist agitation but it can also carry it further and increase its intensity... Contrastingly, the evidence suggests that the inequality in income has, in fact, declined in coastal Andhra" (ch.2: 119.). But, over this period, the affluent farming sections from the rural coastal Andhra have been increasingly shifting to non-farm occupations by leasing out their land, thanks to their command over capital and proliferation of lucrative non-farm opportunities. On all accounts, the area under tenancy has increased phenomenally in coastal Andhra in the recent period. This out-migration could well have led to a decline in rural inequality there, while at the same time contributing to the rise in inequality in areas of their destination, including Telangana and Rayalaseema.

Economic Viability of Telangana State

Viability of a state is usually assessed in terms of its area, population, size of GSDP and public revenues. Area and population assume particular relevance from the governance angle, whereas size of GSDP and pubic revenues have a bearing on economic viability. Tax and non-tax revenues are largely determined by the size of GSDP. In the early years of development with low levels of per capita GSDP, the compulsions of a critical minimum of public revenues for maintenance of general administration, police, judiciary and other essential services did not, in general, favour smaller states. In fact, at the time of the formation of Andhra Pradesh, a strong consideration favouring integration of Telangana with the Seemandhra was the persisting revenue deficit of Seemandhra and existence of revenue surplus for Telangana. Over the last half a century, however, with the steady growth in per capita income and tax revenues, economic viability in terms of public revenues is no longer a binding constraint in many cases. Therefore, the Srikrishna Committee has done well by focusing on GSDP and not getting into the issue of public revenues (except for a passing mention in the Report) while discussing viability.

The Committee observed that "Telangana region (excluding Hyderabad) ranks 15th in the list of 28 states (excluding Andhra Pradesh) in terms of the absolute amount of GDP, and is listed above the states of Chhattisgarh, Jharkhand, Himachal Pradesh, Goa and all the North eastern states (Appendix 2.25). In terms of per capita income, Telangana (excluding Hyderabad) ranks 13th in GSDP as well as in per capita income... Thus, from the point of view of sheer size of economy, Telangana as a new state can sustain itself both with and without Hyderabad" (ch.2: 121).

Having said this, the Committee duly noted the apprehensions of Seemandhra concerning its own economic prospects and viability in case of statehood for Telangana is conceded: fear of local entry taxes and cess scuttling free trade and enhancing cost of business and increasing prices of goods and services; local laws constraining physical movement of goods and services between neighbouring regions and states; and losing a major market inherent in huge population, business and market concentration of the city of Hyderabad (p.122).

The Committee after simply observing that "on this count, separation of AP can be a negative factor which inhibits economic growth of newly formed states" (p.122) goes on to suggest that "it is important to keep the city/district/urban agglomeration of Hyderabad accessible to people and businesses from any of the regions of AP or for that matter from any part of India and abroad. This can be accomplished irrespective of whether the political control over Hyderabad is wielded through a united AP or otherwise" (p.123).

As it is, we have a common national market in the country in which people belonging to several regions and states participate for mutual advantage. Restrictive practices in trade and business equally hurt those practicing them and, in any case, the Union Government cannot remain a mute spectator to such practices. There can hardly be any basis for these fears when there is no evidence of restrictive practices in trade and business between the existing states in the country, including those recently created.

Social Inclusion

The Srikrishna Committee Report contains an illuminating analysis of the issues bearing on social inclusion, based on the data assembled by the Committee from the authentic sources on several indicators of development according to regions, castes, religion and gender, supplemented by the extensive field visits and interviews by the members of the Committee (ch.7). An important fact emerging from these data—not widely known so far—is that the upper or high castes account for as high as 32 per cent of total population in coastal Andhra and 24.2 per cent in Rayalaseema, as against a low level of 10.7 per cent in Telangana (including Hyderabad). Thus the socially disadvantaged sections comprising SCs, STs, Muslims, other minorities and other backward classes (OBCs) constitute as high as nearly 90 per cent of total population in Telangana, a little over three-fourths and two-thirds of total population in Rayalaseema and coastal Andhra (Table 7.1, p.360).

Since the upper castes would, in all probability, predominate among the migrants from Seemandhra into Telangana, this may be contributing to reinforcing the existing inequalities in Telanagana, characterised by feudal antecedents. It is pertinent to note, at this stage, the following observations of the Committee:

> The Upper castes in Rayalaseema and coastal Andhra are vehemently against the idea of dividing the state; their greatest fear being the loss of Hyderabad. The accommodation between these two regions has been in terms of political domination by Rayalaseema and economic domination by coastal Andhra. Together the two regions have ruled the state through Congress and TDP political formations. Telangana feels dominated by the upper castes of these regions and its struggle is primarily to shake off their yoke. (p.390)

According to the Committee, Telangana separatists argued before it that because of its higher proportion of disadvantaged social groups, Telangana will be a more socially, economically and politically equitable and inclusive state (p.389). Dalit groups from all the three regions, citing the viewpoint of Dr B.R. Ambedkar, argued before the Committee that splitting of the state would break the domination of upper castes and provide opportunities for the SC, ST, and OBC communities in both the states (p.368). But the Committee tries to counter this argument by saying that the division of the state may reduce the proportion of the disadvantaged groups in Seemandhra when compared to the unified state. They could thus lose out in Seemandhra (p.389). The Committee overlooks that even after the division, the socially disadvantaged groups would constitute over 70 per cent of the population in Seemandhra. The basic point, however, is that given their large numbers, the weaker sections would be better able to articulate their problems and assert themselves in smaller and relatively homogeneous states because of their shared history and easy communicability (Ambedkar, 1979; Rao, 2010).

Telangana groups argued before the Committee that political power is necessary for channeling resources and development benefits towards the region, that political domination by Seemandhra has adversely affected their prospects and that the only way out of this domination is to have a separate state. The Committee examined this argument by merely counting the number of years for which the posts of Chief Minister (CM), Deputy CM, and important portfolios were held by representatives from different regions, and observed that Telangana side cannot claim total lack of representation as it held important portfolios of home, finance, revenue and irrigation for fairly long periods (pp.406-9). However, the matter is not so simple, as, even when the position of CM was held by 'strong' persons from Telangana, plenty of anecdotal evidence suggests that they felt helpless on vital issues and had to bow down to the wishes of Seemandhra. Political power for different regions flows basically from their respective numbers in the state legislature and not so much by the type of portfolios held or by the time period for which they were held. The latter are in fact determined by the former.

The following observations of the Committee are telling on how Telangana, as conceived by its protagonists, can be socially inclusive, modern, progressive and even trend-setting:

> The Telangana movement can be interpreted as a desire for greater democracy and empowerment within a political unit. As stated earlier, sub-regionalism is a movement which is not necessarily primordial but is essentially modern—in the direction of a balanced and equitable modernisation. Our analysis shows that cutting across caste, religion, gender and other divisions, the Telangana movement brings a focus on the development of the region as a whole, a focus on rights and access to regional resources and further, it pitches for a rights-based development perspective whereby groups and communities put forth their agendas within a larger vision of equitable development. There are strong indications that if Telangana does become a separate state, a movement for separation is likely to follow in Rayalaseema which remains the most backward region in the state (p.413).

When there are such strong grounds for constituting separate Telangana state, it is not fair to withhold its formation simply for fear of its consequences for other regions in the country. Rayalaseema is backward, but its elite—political, bureaucratic and professional—is much stronger than its Telangana counterpart, thanks to its shared history with coastal Andhra and indeed with the rest of the South over a long period in the struggle for freedom, exposure to modern education and progressive land tenures when compared to Telangana. Rayalaseema can be assertive and succeed in protecting its rights and safeguards in the truncated and less heterogeneous Seemandhra after the separation of Telangana. But if it too fails in this and separation becomes inevitable, then this need not be a cause for alarm. The country, in any case, is in for an era of smaller states.

Hyderabad Metropolis

The two major issues of concern for Seemandhra in the event of creation of Telangana state with Hyderabad as its capital are: (a) the possible impact of its separation from Hyderabad on Seemandhra's economy; (b) the prospects for education and employment of youth in Hyderabad; and security of investments, properties, jobs and livelihoods of those from Seemandhra settled in Hyderabad and other districts of Telangana.

The growth of Hyderabad is marked broadly by three phases. The first phase ends with the formation of Andhra Pradesh when the new state inherited the city which ranked fifth and continues with this rank even to this day, among the major cities in the country with its unique location and a highly developed infrastructure built during the Nizam's rule. In fact, this was a major attraction and determining factor behind the integration of Seemandhra with Telangana. It may be recalled that Marathwada and Karnatak which were constituents of the Hyderabad state for a long period contributed, in no small measure, to the development of Hyderabad city but had no claims to it because of its organic link with Telangana. Over the second phase stretching up to the start of globalisation, Hyderabad continued to grow linked basically with the regional economy. In the third phase covering the period of globalisation over the last two decades, the character of the city underwent a major change with its phenomenal growth and increased links with the national and global economy which, as brought out by the Srikrishna Committee, now predominate over its links with the state economy.

This on-going third phase is of special interest. A striking feature brought out by the Committee is the concentration of information technology (IT) and information technology enabled services (ITES) almost solely in Hyderabad accounting for 99 per cent of the total exports from the state (ch.6: 315). While there are local investors in IT and ITES sectors, the composition of the firms established in the city is largely national and international in character. Of the 94 groups identified separately for investments in Hyderabad and Ranga Reddy, 74 are from outside Andhra Pradesh, testifying to the breadth of investment linkages in the region. Even in construction and real estate, finance and business services, it appears that private Andhra Pradesh based investors are not dominant. As such, the industry is connected more to the national (through investment) and global economy (through the market) than it is to the regional economy. Its private investments would be driven by the overall climate of certainty and availability of infrastructure, labour and other services (pp.316, 330, 332).

As regards labour, the traditional out-migrating states of Eastern India, Rajasthan and Uttar Pradesh, which accounted for 23 per cent of the migrants earlier, are now contributing 40 per cent. Thus, according to the Committee, Hyderabad's pattern is beginning to look more like Mumbai's. This is consistent with Hyderabad's greater integration with national and international markets and its growth as a national and global city (pp.323-4).

The Committee's narrative of key cities in Andhra Pradesh state suggests that the economic interdependence between them may be limited. Hyderabad and each urban centre in coastal Andhra and Rayalaseema appear to have their own economic base hinterland and growth drivers. While Hyderabad is much larger than the other cities in the state, it appears that other cities across Andhra Pradesh do not seem to be solely dependent on it for market linkages and other services (pp.336-7).

It is clear from the Committee's findings that the adverse impact, if any, on the economy of Seemandhra on account of separation of Telangana with Hyderabad as its capital is not going to be significant.

The second issue, *viz.*, the protection of the interests of Seemandhra migrants in Hyderabad and other districts of Telangana is going to be extremely important and challenging. As the Srikrishna Committee points out, migrants from the three regions, especially from coastal Andhra, have contributed substantially to the economic growth of the city and continue to hold a stake in important businesses (p.338).

Access to educational institutions and employment in Hyderabad is important to youth from all the three regions of the state. Small trade persons and semi-skilled workers from all over the state also find a niche in the city. People from the three regions have developed strong material and emotional attachment to the metropolis and fear loss of access in case of changes in the state's contours (p.339). In view of this, the Committee's suggestions in the second best option (option 5) regarding the protection of the interests of Seemandhra migrants in Hyderabad and other districts of Telangana deserve serious consideration.

Summary and Conclusion

According to the Srikrishna Committee, Telangana has shown growth dynamism but income inequalities have been increasing. There are also continuing concerns about irrigation, education and employment. Telangana can be economically viable as a separate state with or without Hyderabad. Telangana would also be socially inclusive, facilitating equitable development. Statehood for Telangana is demanded by a large majority of people in the region cutting across caste, religion and gender. Hyderabad metropolis has acquired a growth momentum of its own, driven largely by investments from the rest of the country and serving the national and global market. Hyderabad becoming the capital of Telangana may not adversely affect the economy of Seemandhra.

Given these findings, it appears the Srikrishna Committee has very narrowly missed recommending statehood for Telangana with Hyderabad as its capital, as the 'first best' option. Scales of justice got tilted against Telangana on account of vehement opposition from Seemandhra and Committee's own fears about the demands for smaller states cropping up as well as Telangana becoming a stronghold for Naxalism and Religious Fundamentalism. Above all, it shows that in our system opting for a radical change by moving away from *status quo* is not easy.

Of all these fears, the concerns of Seemandhra, especially those centering on the prospects for its youth in Hyderabad, are most legitimate. Addressing these concerns seriously is critical to the resolution of the crisis so as to pave the way for the formation of Telangana state. This calls for a constructive dialogue between the leaders of both the regions with understanding, spirit of accommodation and statesmanship.

References

Ambedkar, Dr Babasaheb (1979). *Writings and Speeches*, Vol. I (Part II - On Linguistic States). Mumbai: Education Department, Government of Maharashtra.

Galab, S., E. Revathi, and Prudhvikar Reddy (2009). "Farmers' Suicides and Unfolding Agrarian Crisis in Andhra Pradesh", in D. Narasimha Reddy and Srijit Mishra (eds.), *Agrarian Crisis in India*. New Delhi: Oxford University Press. pp. 164-99.

Rao, C.H.H. (2010). *Regional Disparities, Smaller States and Statehood for Telangana*. New Delhi: Academic Foundation.

Rao, S. Kishan and Rahul A. Shastry (2009). *Andhra Pradesh Economy-Dynamics of Transformation with a Focus on Regional Disparities*. Hyderabad: National Academy of Development.

Economic and Political Weekly XLVI(5), January 29-February 4, 2011

25

Hundred Years of Telangana

I am delighted to be in Palamur (Mahboobnagar) again after two decades. In 1991, I and other members of the National Commission on Rural Labour had visited this district to gain fist-hand knowledge about Palamur labour known for their severe poverty and migration. Mahboobnagar continues to be the most backward district in Andhra Pradesh, and labour distress and its out-migration on a large-scale continues.

According to the Human Development Report prepared recently by the Centre for Economic and Social Studies (CESS), Hyderabad, Mahboobnagar district ranks 23rd or lowest among all the districts in Andhra Pradesh in terms of Human Development, that is, when literacy rate and infant mortality rate are taken into account along with per capita gross state domestic product (GSDP). Ironically, Mahboobnagar is adjacent to Hyderabad—the most developed district in the state. This coexistence of the most backward and the most developed districts bordering on each other is a puzzle which would have existed for the last 100 years. As such, this poses a big challenge for study and analysis.

All the same, we find that in the post-economic reform period of the last 15-20 years, Mahboobnagar district experienced a high growth rate in GSDP of over 8 per cent per annum. In fact this growth rate is highest among all the districts of the state. But the level of per capita income of this district continues to be low—though slightly higher than that for Srikakulam and Vijayanagaram districts for a few years. Even so, this high growth rate of the district poses another intellectual challenge for enquiry. Is this due to the impact of the adjoining Hyderabad metropolis which has experienced a phenomenal growth in the post-reform period? What are the economic linkages between these two districts? How are the benefits from such a high growth in

Mahboobnagar district being shared between different social groups? In the light of post-reform experience elsewhere in the country, in all probability, income inequalities must have been increasing in this district with very little benefit trickling down to the poor.

Out of the last 100 years of Telangana, I myself witnessed the events for at least 75 years, starting from my days at the primary school in my native village, Marupaka, in Karimnagar district, where I was admitted when I was 6-7 years old. I distinctly remember the Silver Jubilee celebrations of the last or 7th Nizam of Hyderabad when we were made to wear the Silver Jubilee badges with pride. The Nizam with his *Dastar* was ingrained in these badges made from pure silver, sold for one rupee coin each.

I was a witness to feudalism in operation in Telangana those days within a society and polity of multiple layers of domination—the Nizam under the British, the *Jagirdars*, Deshmukhs and Patel-*Patwaris* under the Nizam and the common people groaning under their yoke. It was worse still for Mahboobnagar district. Here you had an additional layer of domination, i.e., '*Samasthans*' or petty states within the Nizam state. This together with highly unfavourable agro-climatic conditions of the district probably explain its extreme backwardness. '*Vettichakiri*' or forced labour without payment of wages and other kinds of extortions were rampant in the villages of Telangana. Mahboobnagar presents yet another puzzle: despite extreme backwardness and poverty, extremist activities were not seen here as in other parts of Telangana. This could well be due to low staying power of the poor and their large-scale out-migration. But there could also be certain historical and cultural factors explaining this phenomenon. All these need investigation.

Under these circumstances, library movement emerged in Telangana led notably by the Andhra Mahasabha for awakening the people. I myself happened to be a young librarian in-charge of the library at the Andhra Mahasabha office in Karimnagar in the mid-1940s. At the dawn of this library culture, I remember it was difficult to persuade people to return the books they borrowed. Quite a few considered it an affront to their dignity to be reminded about the return of the books! There were no civil liberties worth the name in the state. The activities of the All Hyderabad Students' Union of which I happened to be the secretary in Karimnagar were branded as 'political' and some of us were asked to resign from the Union and apologise. Upon refusal, we were rusticated from the school in June 1946, by withholding the results of our exam for the 9th standard. Thanks to Urdu being the medium of instruction, I could pass High School Examination (HSE) from the Aligarh Muslim University, as a private candidate, the same year (1947).

Then came the Police Action in 1948 and integration of the Hyderabad state with the Indian Union. Soon after, the abolition of *Jagirdari* and *Zamindari* and enactment of tenancy legislation followed. But, following the agrarian struggle of 1940s in Telangana and the progressive land reform legislation enacted after the Police Action, not enough time was left for basic social transformation in the region before it was integrated with Andhra Pradesh in 1956. Incidentally, I recall that just before the formation of Andhra Pradesh in 1956, I had an opportunity of visiting Mahboobnagar as the General Secretary of the All Hyderabad Students' Union to meet and discuss with the students, among other things, the issue of integration of Telangana with Seemandhra.

After the merger of Telangana with Seemandhra, contrary to the expectations of the votaries of this merger, including myself, it has been a long story of discrimination against Telangana for over half a century in respect of the allocation of public investment, water resources, education and employment. Starting from a low base, Telangana did experience a 'catching-up' process by showing some growth despite several constraints without which it would have experienced much higher growth. But inequalities in income have been increasing with mounting unrest and discontent among the people of Telangana against discrimination and dominance by the business and political elite of Seemandhra. As a result, we are seeing an unprecedented movement for separate Telangana State—far more widespread or rather universal when compared to the agitation of 1969.

On the basic issues bearing on the formation of a separate state of Telangana, *viz*., the reasons for discontent in Telangana, its economic viability and wishes of the people of Telangana, the Committee headed by Justice Srikrishna, appointed recently by the Government of India to report on the situation in Andhra Pradesh, found them strongly favouring the formation of a separate Telangana state with Hyderabad as its capital. Given its own findings, it appears the Srikrishna Committee has very narrowly missed recommending statehood for Telangana with Hyderabad as its capital, as the 'first best' option (see the comment above on the Report of this Committee).

In the meantime, in response to a petition filed by Shri M. Narayan Reddy, ex-Member of Parliament (MP), the Andhra Pradesh High Court gave a judgment on 23 March, 2011, against withholding chapter 8 of the Report of this Committee by observing that "withholding of note from public is arbitrary, unreasonable and unconstitutional, and that a meaningful discussion, with reference to the Report submitted by the Committee, cannot take place, unless it is made public". The Court observed further that "the objective in preparing a separate note and delivering it to the respondent was more an effort to persuade the Union of India to desist from showing any inclination towards Option No.5, i.e., formation of Telangana State." The Court also opined that the Committee travelled beyond the terms of reference in its endeavour to persuade the Union of India not to concede to the demand for Telangana by offering its detailed advice on "Political Management" and "Media Management" towards this end, in a three-page supplementary note appended to Chapter 8.

According to the Srikrishna Committee, the option of creating Telangana state (option 5) is to be exercised only if the decision can be reached amicably amongst all the three regions by instilling confidence in the people of coastal Andhra, Rayalaseema and others who have settled in Hyderabad and other districts of Telangana. (ch.9: 450-8).

Compare this with what the States' Reorganisation Commission (SRC), headed by Justice Fazal Ali, said in 1956. They were democratic to the core. They said Telangana should be merged with Andhra after 5 years only if two-thirds of the legislators from Telangana opted for it. They did not talk about the opinion of Andhra legislators.

The present approach of insisting on consensus among all the regions of Andhra Pradesh betrays a lack of democratic spirit in so far as it ignores the universal demand for statehood in the Telangana region. It also shows scant respect for the constitutional provisions which do not require the consent of state legislature in this regard, as well as lack of faith in the people of Telangana by invoking the fear of Maoism and religious fundamentalism. This approach ignores the history and traditions of Telangana, proverbial for its cosmopolitanism and religious harmony. It is here that people like Makhdoom Mohioddeen and Shoiebulla Khan lived and died for religious harmony. It is here that an armed struggle was launched during the 1940s against feudalism for securing land to the tiller when thousands of lives were lost but was withdrawn before the first general elections when the opportunities for democratic and peaceful movements were opened up.

Unfortunately, however, our democratic system has not been responding adequately to the aspirations of the common people. The functioning of our democracy is being twisted and moulded by the narrow-interest groups. Backward regions and weaker sections are becoming victims in this process. It is true that 'ideas' are stronger than 'interests' but 'interests' seem to be generating their own ideas and even so-called 'ideals.' Some of those who often talk of 'values' of democracy seem to forget that the foremost value in a democracy is the 'democratic spirit' which means, among other things, respect for the wishes of the people expressed through the democratic process. It is surprising and distressing to see even intellectuals trained in the first principles of democracy not showing respect for democratic aspirations of the people. Therefore, how to make our democratic system responsive to the legitimate aspirations of the people and become a catalyst for social transformation is the foremost challenge.

The leaders and builders of Telangana of the future should not forget this lesson learnt the hard way. Telangana that we are struggling to create has to be the one where the elected elite is truly responsive to the democratic aspirations of the people. It has to be a vibrant and socially inclusive Telangana with a participative democracy and not of the elite representing vested interests. This is perhaps the most important lesson learnt at the culmination of 100 years of Telangana.

For obvious reasons discussed above, Mahboobnagar district has to be top on the development agenda of the Telangana state. The prospects for development of this district in the new smaller and more homogenous state are bright. The possible spin-off effects of contiguous Hyderabad metropolis on the development of the district, already in evidence as mentioned earlier, should be fully explored and exploited. These are going to be very significant, especially because the importance of the services sector will increase in the ensuing phase of development. Further, the proximity of this district with the capital city should ensure more informed policy attention as well as better monitoring of performance.

Keynote Address at the seminar on "Hundred Years of Telangana" orgainsed by the Palamuru Adhyayana Vedika, Mahboobnagar, February 26-27, 2011.

26

Political Empowerment of Telangana

Introduction

Achieving economic growth with social justice has always been the foremost objective of economic planning in India since independence. Although the achievements in this respect have been substantial, growth achieved was much slower than planned and reduction of poverty was also slow. It was after the launching of economic reforms in the early 1990s that the growth rate of the economy accelerated, but the reduction of poverty became even slower with rising inequalities in income. In view of this, the plan strategy was reviewed and the Eleventh Five Year Plan aimed at achieving faster and inclusive growth. In the last few years, however, issues like the degradation of environment and large-scale displacement of people and destruction of their livelihoods, especially in tribal areas, on account of indiscriminate mining and certain other projects have come into sharp focus. Responding to these growing concerns, the Government of India has recently come out with the threefold strategy of faster, inclusive and sustainable development during the Twelfth Five Year Plan.

But the effective implementation of such a strategy calls for vastly improved governance at the state level, social cohesion and proper accountability of administration with the active participation of the people down to the grass-roots level. Experience has shown that these tasks can be better accomplished in smaller states.

Increasing Role of the State in the Post-Reform Period

Liberalisation and globalisation have, no doubt, released the creative energies of our entrepreneurs and people at large. But, contrary to conventional wisdom, the role of the government has become more important in the post-economic reform period even after the removal of licensing, controls and other restrictions which led to the widening of economic freedom for industry, commerce and trade. This is because the state continues to be the main propeller for the provision of infrastructure like irrigation, power, roads, transport and financial institutions either directly or indirectly through the private sector or through public-private participation.

In general, despite such government intervention for infrastructure development, the impact has been towards increasing inequalities between different states and between different regions and income groups within the states, particularly among the larger ones. This is because official patronage tends to favour the better-off regions and income groups already well-endowed with resources, skills, power and influence. This experience clearly shows that certain backward regions, if constituted as separate states, may use this enormous potential offered by state power effectively for their development.

Better Performance of Smaller States

This is amply borne out by the recent experience with the creation of smaller states like Uttarakhand, Chattisgarh and Jharkhand. These states have been able to achieve much higher growth rates in their gross state domestic product (GSDP) than the targets set for them by the Planning Commission. In fact their growth rates have been significantly higher than those achieved by their parent states, namely Uttar Pradesh, Madhya Pradesh and Bihar. Better governance in these smaller states has contributed to attracting greater private investment from outside as well as to better planning and utilisation of resources. This experience shows that the political commitment necessary for a focused attention on the problems of growth, equity and sustainability can be better ensured in smaller states which are relatively homogeneous.

An interesting consequence of carving out such smaller states is that their parent states become smaller and, for similar reasons of viability and efficiency, they can be expected to grow faster in course of time. An extremely encouraging development in this respect relates to Bihar where its GSDP growth rate has accelerated in the last few years. This may be attributed to improved governance, of late, in this state because, among other things, with the creation of Jharkhand, Bihar has become more cohesive and much smaller in area with the size of its population getting reduced by one-fourth.

Among the general category states in the country, the growth performance of smaller states has been particularly impressive in the post-reform period. During the pre-reform period of 1980s, there was hardly any correlation between the size of the state measured by population and growth rate in GSDP. Over the last decade, however, one finds the emergence of a significant correlation between the size of the state and growth rate achieved, smaller states such as Punjab, Haryana, Gujarat, Karnataka, Kerala ,Tamil Nadu, Chattisgarh and Jharkhand showing much higher growth rates in GSDP than the larger ones like Uttar Pradesh, Bihar, Madhya Pradesh and Rajasthan. This experience shows that smaller states have been better placed to exploit the opportunities opened up by economic reforms. Several of these smaller states also show higher levels of per capita GSDP as well as higher tax-GSDP ratios indicating their economic viability.

Politics of Development in Bigger States: Discrimination Against Backward Regions

Maharashtra and Andhra Pradesh stand out as glaring cases of politically dominant groups hailing from the developed regions discriminating against the backward regions in respect of investment and employment. Maharashtra, because of its developed physical and financial infrastructure, has been able to attract disproportionately large investments in the post-reform period whether from the all-India financial institutions or foreign direct investment (FDI). But these investments have been concentrated basically in the developed pockets like the Pune-Nasik belt to the neglect of backward regions like Vidarbha and Marathwada. This has happened despite the constitutional provisions in respect of Maharashtra, unlike for Andhra Pradesh, to safeguard the interests of backward regions through the establishment of Regional Development Boards with special powers to the Governor for monitoring the progress. Clearly, politics of development have had an overriding influence undermining the constitutional provisions in favour of backward regions.

As for Andhra Pradesh, I would like to recall first what I once asked late Professor K. Jayashankar: "What explains so much discontent and unrest and vehemence for statehood in Telangana when compared to Vidarbha which is as backward or even more backward than Telangana?" His quick and crisp response was quite revealing which answered several other unasked questions: "Vidarbha has been neglected whereas Telangana has been exploited!" By now, there is plenty of evidence to show how Telangana has been deprived of its legitimate share in public investment and employment and how the natural resources of the region like land, water, forests and minerals have been alienated on a large scale to the enterprising migrants from Seemandhra. Private investments from within the country and abroad are welcome in so far as they create productive assets and jobs for the locals. But investment flows from the developed to the less developed regions within a state often result in the appropriation of jobs by the migrants, apart from the alienation of natural resources. Whereas such investments do result in stepping up the GSDP growth rate of the region, deprivation of jobs and natural resources and degradation of environment lead to discontent, unrest and tensions.

On the question of various deprivations suffered by the people of Telangana, I cannot do better than refer to the illuminating and effective presentation by the Telangana Chambers of Commerce and Industry in Support of Separate Telangana State, made before the Committee for Consultation on the Situation in Andhra Pradesh, headed by Justice Srikrishna. According to this submission, among the category of medium and large-scale industries in the state, 62 per cent of companies were located in Telangana during the period 1991-2010. Out of those located in Telangana, as many as 75 per cent were owned by Andhra entrepreneurs; the remaining 25 per cent owned by those from Telangana. While companies located in Telangana accounted for 56 per cent of total employment generated, what is revealing is that as much as 75 per cent of this employment generated in Telangana belonged to Andhra youth, remaining 25 per cent to the Telangana youth. As regards Small-Scale Industries till 2007, 56 per cent of total employment was generated in Telangana. Out of this 70 per cent belonged to Andhra youth, remaining 30 per cent for Telangana youth.

A case study of Hyderabad Unit of Bharat Heavy Electricals Limited reveals that in the category of General Managers and Executives, 80-90 per cent belonged to Seemandhra. Even among the contract labour nearly two-thirds, and among staff and workers a little over half belonged to Seemandhra. As for the software industry in Hyderabad, the local youth do not constitute even 5 per cent of total employment.

The above narration explains the paradox of high GSDP growth rate experienced in Telangana coexisting with mass discontent and strong demand for separate statehood. What is at issue, therefore, is not whether growth has been taking place, but how the benefits of growth are being shared by different sections of people. Therefore, when some of the protagonists of Telangana asserted that the real issue is 'self-respect' through political empowerment or statehood rather than 'development', what they mean is that the real issue is 'inclusive and sustainable development' which is not possible unless people's 'self-respect' is restored through political empowerment or statehood.

Who wields effective political power at the state level is central to the understanding of regional inequalities within a state. Andhra Pradesh provides a glaring illustration of state power being decisively influenced by the politico-business elite from a particular developed region to sub-serve its interests while denying level playing field to commerce and industry in a backward region like Telangana. The Srikrishna Committee specifically names at least four major business groups from Seemandhra having enormous political clout as members of parliament and ministers who are actively engaged in real estate industry and other business in Telangana (chapter 6: 319). Yet, the Committee did not face squarely the issue of political dominance as central to the statehood demand for Telangana. Otherwise, it is difficult to explain its recommendation for keeping the state united by creating a statutorily empowered Telangana Regional Council—an experiment which miserably failed earlier.

What Next?

Telangana is a national issue reflecting the aspirations of the people in different parts of the country for shaping their destinies through political empowerment. The demand for carving out four smaller states from Uttar Pradesh—the largest state in the country—represents such aspirations. Unless it is believed that the number of states in the country would be frozen at the present level for ever, the creation of Telangana state would be foremost on the agenda, if there is to be any reorganisation of states at all. This is because, as discussed above, the unrest giving rise to the demand for Telangana state is bound to grow as the issues causing this discontent cannot be resolved except through political empowerment. The demands for creation of smaller states in Uttar Pradesh and in other parts of the country only strengthen Telangana's case for statehood, but do not tie it up with them. Rather, the reality of Telangana becoming a state may push the demand for Vidarbha and smaller states in Uttar Pradesh to the centre-stage.

This has to be a continuing process, each case being decided on merits rather than an attempt at once-for-all reorganisation of states in the country on certain theoretical or *a priori* considerations set out by a body like another States' Reorganisation Commission.

Inaugural Address delivered on the occasion of launching of "Telangana Chambers of Commerce and Industry", Hyderabad, 21 December, 2011; Published in *Mainstream*, Vol. L, No. 9, February 18, 2012.

27

The Promise of Telangana

THE decision of the United Progressive Alliance (UPA) and the Congress Working Committee to accede to the long-standing demand for a separate Telangana state from Andhra Pradesh ends uncertainty holding up development and opens up opportunities for harmonious and inclusive development of both the states. But it is necessary to avoid short-term and sectarian considerations while drawing up their boundaries to exploit the long-term opportunities for development provided by the need for building the new capital of Andhra Pradesh.

The idea of including two districts of Rayalaseema, Kurnool and Anantapur, in Telangana is apparently kept open, however slender may be the chances for such a proposal. Like Telangana, Rayalaseema acquired a distinct regional identity of its own because of agro-climatic, historical and political factors. Therefore, any attempt to divide it is bound to meet stiff resistance from the people of Rayalaseema.

Notwithstanding its enterprising elite, Rayalaseema is the most backward region in the present state. Its development requires a concerted plan for the whole of this homogeneous region as well as the necessary institutional arrangements and political clout that are possible only when the four districts of the region are together in one state. One hopes that short-term political advantages arising from the demographics of religion and caste (See *The Contested Ideas of Rayala Telangana*, Economic Times, July 30) will not be allowed to favour a decision that could hamper the development of this backward region and, indeed, of both the new states.

An issue of concern following the creation of Telangana state is the possible impact on the economy of Seemandhra (the residual state of Andhra Pradesh consisting of Rayalassema and Coastal Andhra regions) after its separation from Hyderabad. Another is the prospects for education and employment for Seemandhra youth in Hyderabad and security of investments, properties, jobs and livelihoods of those from Seemandhra settled in Hyderabad and other districts of Telangana.

In the current phase of globalisation, the character of Hyderabad underwent a major change with its phenomenal growth and increased links with the national and global economy that, as brought out by the Srikrishna Committee on Andhra Pradesh, now predominate over its links with the state economy.

A striking feature brought out by the committee is the concentration of IT and ITeS almost solely in Hyderabad accounting for 99 per cent of these exports from the state. But, of the 94 groups identified separately for investments in Hyderabad and adjoining Ranga Reddy districts, as many as 74 are from outside Andhra Pradesh. Even in construction and real estate, finance and business services, private Andhra-based investors are not dominant.

The committee's narrative of key cities in Andhra suggests that the economic interdependence between them may be limited. Hyderabad and each urban centre in coastal Andhra and Rayalaseema have their own economic hinterland and growth drivers. While Hyderabad is much larger than other cities in the state, the latter are not solely dependent on Hyderabad for market linkages and other services. Thus, the adverse impact, if any, on the economy of Seemandhra on account of separation of Telangana with Hyderabad as its capital is not going to be significant.

The second issue is the protection of the interests of Seemandhra migrants, especially youth, for education and employment in Hyderabad. The proposed measures to protect such interests could be similar to the decision to retain Hyderabad as joint capital for 10 years. In any case, the issue would call for a dialogue between leaders across regions with a spirit of accommodation.

Construction of a new world-class capital for Seemandhra holds the greatest promise, among other things, for easing pressure on education and employment of the youth. India is on the threshold of rapid urbanisation. Construction of new capital for Andhra could be a major source for reviving growth and increasing employment. It also provides an opportunity for constructing an eco-friendly, energy-efficient and slum-free city with public transport and safeguards for easy mobility for pedestrians.

Would the people of Telangana be hostile to the businesses in Hyderabad owned by people from coastal Andhra? Several business persons in Hyderabad hailing from Andhra interviewed immediately after the decision to bifurcate the state reportedly expressed relief over the ending of "uncertainty". Hardly any apprehensions were ventilated.

The current regional conflicts relate to the distribution of benefits from public investments and employment. Private enterprise was unaffected by separatist agitations. Conflicts of interest arise when state power is used by the politico-business elite from the Andhra region to deny a level playing field to commerce and industry in Telangana.

Economic Times, 2 August, 2013.

28

The State of Telangana

A Perspective for Inclusive and Sustainable Development

Introduction

The formation of the new State of Telangana promises to address and fulfill the long cherished hopes and aspirations of over 35 million people. The merger of the former Hyderabad State with the Indian Union in 1948 marked the end of feudalism and opened up a vista of opportunities for development for the people of Telangana. But, not much time was allowed for social change and transformation in Telangana before it was merged with the Andhra region to constitute the State of Andhra Pradesh in 1956. Telangana's merger with the more resourceful, educated, skilled and politically dominant Seemandhra, far from releasing local initiative and enterprise, gave rise to new tensions and universal discontent among the people of Telangana consisting largely of weaker or disadvantaged sections, e.g., SCs, STs, BCs and minorities, who constitute nearly 90 per cent of the total population.

With the creation of new State of Telangana, there would be high expectations from the common people for their economic betterment and for opportunities to participate in shaping their destinies. Nothing short of a new social framework that allows inclusive and participatory development would be able to meet these aspirations. This is desirable as well as workable. A sociopolitical arrangement that allows the common people to share political power and responsibilities at various levels would enable them to articulate their real problems and explore workable solutions in keeping with local resource endowments. Such an arrangement would also enable them to see the possibilities and limitations of development with the available resources and can elicit constructive effort from them ensuring stability and social harmony.

Inclusiveness could not be achieved so far in the bigger state of Andhra Pradesh because the voice of the disadvantaged sections remained fragmented. Experience shows that the traditionally entrenched interests are perpetuated in bigger and heterogeneous states because of their easy connectivity arising from their access to large resources, power and influence. The weaker sections, on the other hand, can come together, organise themselves and raise their voice effectively in a relatively homogeneous state because of common history and traditions, and hence easy communicability.

Social Composition of Population

Tribal population is the most disadvantaged section socially and economically with negligible political voice. They live in remote areas and are subjected to land alienation on a large scale. Hardly any initiative was taken in Andhra Pradesh to restore their lands despite the strong recommendations made by a High-Level Committee headed by Koneri Ranga Rao, a Minister, constituted by the government. There, the administration is alienated from the people and so the area has been a breeding ground for extremist activities. This has been treated not as a socioeconomic issue but mainly as a 'law and order' problem. Because of this, the plight of the Girijans has been perpetuated and the extremist activities have been surfacing time and again, notwithstanding the claims of 'success' in this regard by the authorities.

According to 2011 Census, ST population constituted around 9.3 per cent in Telangana as against 5.3 per cent in the residual State of Andhra Pradesh. Thus, as much as 60 per cent of the ST population of undivided Andhra Pradesh State is concentrated in Telangana. Their voice can be expected to be more effective in the Telangana Sate, not the least because their representation in the State legislature and other elected bodies at different levels would be proportionately greater. SC's account for about 15.4 per cent of population in Telangana State as against 17 per cent in the residual State of Andhra Pradesh.

The population of Muslims was as high as 12.5 per cent according to 2001 Census in Telangana when compared to 6.9 per cent in Seemandhra. As many as 61 per cent of Muslims of undivided Andhra

Pradesh live in Telangana, of whom 60 per cent are spread over in different districts other than Hyderabad. They too can be expected to have greater political clout in the Telangana State in determining their fortunes as they can more easily relate themselves with the rest of the disadvantaged sections of the society in the struggle for a better and more secure livelihood. Social harmony between people professing different religions and speaking different languages has been proverbial in Telangana because of their shared history and traditions spanning over centuries (Rao, 2010).

Radical land reforms were the prime agenda for the peasant movement in Telangana in the 1940s. However, not enough time was available for this process of agrarian reforms and social transformation to run its course. In fact, it was interrupted and there was some retrogression following the integration of Telangana with the Andhra region. In a larger and heterogeneous state like Andhra Pradesh there was neither adequate perception of this problem nor willingness to address it by the dominant political leadership which hailed basically from the developed region.

Thus the weaker sections constituting the large majority of population in the Telangana State and, for that matter, in the residual State of Andhra Pradesh would be better able to articulate their problems and politically assert themselves in smaller and relatively homogeneous states. The creation of Telangana State would thus strengthen the forces of social inclusion in both the states.

Feasibility of Inclusive Governance

The population of Telangana State is over 35 million now—much more than 30 million for the whole of Andhra Pradesh, including Telangana, at the time of its formation in 1956. The demands on governance have multiplied over this half a century. Apart from commitment to the development of the region, a smaller state being more easily accessible to the common people can intelligently and speedily grapple with their problems.

Moreover, governance at the grass roots can be improved by ensuring greater accountability of performance through the panchayati raj institutions (PRIs). But the devolution of functions, finances and functionaries to these institutions in the undivided state was far from adequate. The devolution index, constructed by including the above parameters, was much lower for Andhra Pradesh (50.10) when compared to the other southern states, e.g., Kerala (75.00), Karnataka (69.50) and Tamil Nadu (67.10) (Reddy, 2012). It is indeed ironical that in the undivided Andhra Pradesh, those in power owing allegiance to Rajiv Gandhi, who visualised 73rd and 74th Amendments to the Constitution, did not take any initiative to strengthen these institutions. On the contrary, attempts were made to undermine them by floating several top-down schemes and parallel implementation structures, even naming some of these schemes after Rajiv Gandhi! In smaller and relatively homogeneous states of Telangana and Andhra Pradesh, the empowerment of these local elected institutions can be expected to be high on the agenda, not the least because of the greater pressures these elected representatives can bring to bear on the new establishments.

The above social framework is a pre-requisite for setting up an appropriate agenda for inclusive development and for the effective implementation of programmes by ensuring accountability to stakeholders at different levels (Dawra, 2012). The key areas of development which suffered neglect and severe policy distortions so far and hence need to be accorded high priority in the Telangana State are: land issues; irrigation and power; and education and health.

Land Issues

The primary objective of the first round of land reform legislation in the 1950s and even of the second round in 1970s was to 'secure' land to the tiller. The issue now is basically to 'restore' to the original owner-tillers land alienated illegally, apart from distribution of land to the landless poor, to the extent feasible. Conserving the available public land and augmenting it further for essential public purposes such as schools, hospitals, house sites for the poor, etc., in the wake of rapid urbanisation is another issue.

On land issues, the recommendations of the Koneri Ranga Rao Committee, which are quite moderate and concerned basically with more effective implementation of the existing laws (See chapter 23), can be made the basis. These include restoring lands to tribals and checking further alienation of their land by reopening and re-examining of orders in favour of non-tribals and review of large number of cases of illegal occupation by non-tribals; in non-tribal areas restoration of 'assigned' land to the original assignees or, where it is not possible, its resumption by the government for making fresh assignments to the landless poor; and issuing loan eligibility cards to the tenants to enable them to access institutional credit

In the recent period, there has been a massive sale of government land to the private parties and corporations in Andhra Pradesh. The data on the public land still available in Telangana needs to be brought out and assessment be made regarding the extent to which the government would be in a position to secure land for public purposes.

Irrigation and Power

In the last 50 years, there has been a major shift in the pattern of irrigation in Telangana making it more costly for the farmers, highly uncertain and unsustainable: There was a steep decline in area under tank irrigation and slow increase in irrigation under major and medium irrigation projects, leading to the sharp rise in area irrigated through private wells (Telangana Development Forum, 2010).

This process needs to be reversed through a big drive for the repair and renovation of tanks by involving the local people as well as by stepping up public investment for tapping river water for irrigation, especially from Godavari, for which there is considerable potential (Rao, 2006). Already in the 1960s, the Techno-Economic Survey of Andhra Pradesh conducted by the National Council of Applied Economic Research (NCAER) pointed out that "the scope for large scale exploitation of ground water is absent in Telangana as the substrata are trap or granite, incapable of yielding prolific ground water supplies," and that "Godavari water can very well be tapped in future to benefit the region" (Rao, 2010).

Rejuvenation of tanks, wherever possible, can have an immediate beneficial impact on the livelihoods of the common people without much investment (Pingle, 2011). Although the tapping of Godavari water in Telangana may entail substantial use of power for lifting water from the river, there would at the same time be a significant saving of power presently used for well irrigation because of the availability of irrigation water from canals (Vedire and Vedire, 2008).

Telangana is now deficit in power whereas Seemandhra has a surplus. This happened despite the existence of huge unexploited coal resources in Telangana suited to power generation. Some power plants in Seemandhra get the coal transported from Telangana whereas it is far more economical to locate the power plants near the pitheads. The present supply-demand gap for power in Telangana, arising from long-standing neglect and discrimination in the creation of capacity in the region, can be met, in the short-run, only through special measures from the centre.

Given adequate investment from the state as well as the centre, ample power can be generated in Telangana without much time-lag. This would generate large-scale employment locally, including in coal mining, power plants and ancillary activities.

Education and Health

There is a growing demand for education in Telangana, particularly at secondary and higher levels, as a means of employment (Government of India, 2010). This is attributable mainly to low and uncertain incomes from agriculture when compared to the prosperous south coastal Andhra. Surprisingly, this phenomenon noticed way back in the 1960s (Rao, 2010) continues to persist. But the cost of education in Telangana has been rising because of increasing privatisation consequent to inadequate public provision. This is reflected in a rise of annual household consumer expenditure on education as proportion to total annual household consumer expenditure (Telagana Development Forum, 2010).

What is worse, the rise in cost of education has been accompanied by deterioration in its quality. The drop-out rates for students at primary level are quite high in Telangana. Indeed, the increasing demand for education through private institutions is explained basically by the deterioration in the quality of education in government schools and colleges (Centre for Economic and Social Studies, 2008). The major victims of this process of privatisation are students from low income groups, especially the SCs and STs, many of whom can not afford costly education. It has adversely affected their competitive position in the job market (Government of India, 2010). This is reinforced by high spatial inequality, as, except in Hyderabad and Rangareddy districts, the provision of education is inadequate in the rest of Telangana, which continues to lag behind Seemandhra.

The major challenge, therefore, is to ensure affordable and quality education to students at all levels in large parts of Telangana. This requires a significant rise in public expenditure on education. It also calls for greater accountability of school management to the stakeholders through the local elected institutions for the provision of essential services like sanitation, ensuring proper attendance of teachers as well as their periodic training.

Telangana, including Hyderabad, had inherited good health infrastructure from the former Hyderabad state, so that it was ahead of Seemandhra at the time of the formation of Andhra Pradesh. But, as in education, there has been a concentration of facilities in Hyderabad, leaving the rest of Telangana far behind Seemandhra. In terms of hospital beds per lakh population, the gap between Telangana, excluding Hyderabad, and Seemandhra has been increasing. In respect of the number of doctors per lakh of population too, it falls behind Seemandhra.

As in the case of education, there has been a neglect of health services on public account while private sector gained ascendance. Public expenditure on health sector whether as a proportion to total public expenditure or to gross state domestic product (GSDP) has been declining in Andhra Pradesh. Annual per capita government expenditure on health care is now well below per capita household expenditure on health services which accounts for a little over 70 per cent of total per capita expenditure on medical care (Centre for Economic and Social Studies, 2007; Government of India, 2010).

The growth of private and corporate sector in health care has been encouraged by the government in various ways, and lack of government regulation on minimum standards has worsened the situation, especially for the poorer sections of population. Therefore, one of the major tasks confronting the Telangana State is to substantially step up public expenditure on health services, especially on primary health care in rural areas, by making them accountable to the stakeholders. At the same time, there is a need for the regulation of private corporate sector with a view to ensuring minimum standards and for preventing questionable practices.

Conclusions

A new social framework that is participatory and accountable to stakeholders is a pre-requisite for inclusive and sustainable development. Sorting out the pending land issues will provide security for rural livelihoods and the necessary means for raising the incomes of tribal population—the most marginalised section of rural society.

Reducing excessive dependence on well irrigation by expanding surface irrigation through the renovation of tanks and harnessing river waters will contribute immensely to sustainability, apart from reducing farm costs and uncertainty. Development of power on a priority basis is indispensable in the Telangana State for overcoming the inherited shortages, for lifting the river waters for irrigation and to facilitate the growth of manufacturing as well as rural industrialisation in general.

Government should shoulder much greater responsibility towards providing primary and secondary education as well as primary health care by making these accountable to the stakeholders through the local elected institutions and by regulating the private players in these fields.

Restoring land to the tillers, especially in tribal areas, decentralising development by empowering local elected institutions, and stepping up public investment for essential physical and social infrastructure are the three major challenges in the new Telangana State.

References

Centre for Economic and Social Studies (2008). *Human Development Report 2007: Andhra Pradesh*. Hyderabad.

Dawra, Sumita (2012). *Poor but Spirited in Karimnagar—Field Notes of a Civil Servant*. Harper Collins. A joint venture with India Today Group, New Delhi.

Government of India (2010). *Report of the Committee for Consultations on the Situation in Andhra Pradesh* (Chairperson: Justice B.N. Srikrishna). Chapter 3. New Delhi.

Pingle, Gautam (2011). "Irrigation in Telangana: Rise and Fall of Tanks", *Economic and Political Weekly* XLVI(26 & 27). June 25.

Rao, C.H.H. (2010). *Regional Disparities, Smaller States and Statehood for Telangana*. New Delhi: Academic Foundation.

Rao, R. Vidyasagar (2006). *Neellu-Nijalu* or 'Water-Realities'. Hyderabad: Telangana Vidyavantula Vedika-Telangana Development Forum (U.S.A.).

Reddy, M. Gopinath (2012). "Fiscal Transfers to Local Bodies—Andhra Pradesh Experience", Paper presented at the National Seminar on *Centre-State Relations in Indian Fiscal Context* in honour of B.P.R. Vithal. Hyderabad: Centre for Economic and Social Studies. 5-6 November.

Telangana Development Forum (TDF), (2010). *Deprivation to Telangana: Case for Separate Statehood*. Memorandum prepared by the Working Group for submission to Justice Srikrishna Committee, Hyderabad.

Vedire, Venkata Reddy and Sriram Vedire (2008). *Water and Power Crisis: What Makes Telangana State a Must*. Hyderabad.

V

Some Reminiscences

29

National Integration, Poverty and Environment

Relevance of Indira Gandhi

ON the morning of 31st October, 1984, when I was having bath at 3, Safdarjung Lane, New Delhi, where I and my wife were staying, I could spot through the window some unusual running about in the Prime Minister's house, only a few yards away across the lane. Within minutes, our domestic help came running to us breaking the shocking news of her assassination. I was scheduled to meet her soon to discuss the Plan for extending the Green Revolution to the eastern region which she had asked me to prepare in the Planning Commission. Instead, sadly, I had to give a talk at the All India Radio on Indira Gandhi's Contributions to Economic Planning in India.

Her elevation to the Prime Minister's position in mid-1960s owes primarily to the conviction among the then senior leaders of the Congress that such a move would strengthen national unity and national integration: a multi-religious, multi-lingual and multi-regional society like ours needed a person like her as Prime Minister, especially with her family background and credentials.

Emotional integration among diverse groups is possible only when they can experience a sense of belonging to the larger entity. This comes about when every one has a stake in nation-building, giving rise to the feeling that their aspirations can be fulfilled within the system through democratic and peaceful means. The feeling of alienation is bred when there is a sense of neglect and discrimination; and, worse, when there is no hope of redressing through democratic and peaceful means.

A sense of belonging and possibility of redressal and, therefore, confidence in the system was promoted during her tenure by giving a broad-based developmental orientation to the goal of national integration, involving, among others, the far-flung Himalayan and North-eastern regions and the poorer and the under-privileged sections at large in the country.

Recall the events of the mid-1960s when national sovereignty came under severe stress because of 'ship-to-mouth' existence by having to import as much as 10 million tones of food grains annually. She brought the country out of this crisis through her bold decision to usher in the Green Revolution and by extending it in the eastern zone in 1980s, to consolidate food security.

Economic integration of North-Eastern states and the Himalayan region with the rest of the country was achieved through their rapid development by according them the status of Special Category States, whereby 30 per cent of total plan outlay of the country was pre-empted for them. They are increasingly becoming part of the national mainstream. We now find quite a few boys and girls from these regions studying in several colleges and universities in different parts of the country and employed in different departments, especially of the Central Government.

Her another major contribution towards national integration was to align the aspirations of the poor and the under-privileged with the national development effort. Her decision to nationalise commercial banks and to expand banking to the hitherto unbanked areas, particularly rural areas, contributed significantly to raising the saving rate in the economy, and helped to provide resources for innumerable small entrepreneurs, farmers and artisans. Apart from the second round of land reforms undertaken in the early 1970s and programmes for the uplift of small and marginal farmers, a number of poverty alleviation programmes, e.g., Integrated Rural Development Programme (IRDP), National Rural Employment Programme (NREP), Rural Landless Emloyment Guarantee Scheme (RLEGS), Minimum Needs Programme (MNP), etc., were launched from time to time during her long tenure.

Scheduled Caste Component Plan (SCP) and Integrated Tribal Development Projects (ITDP)—which have been in the headlines recently for their severe neglect and diversion of resources meant for them—were started and their performance rigorously monitored during her tenure. As one dealing with these subjects in the Planning Commission, I was a witness to her spending long hours, some times till late in the evenings, reviewing the implementation of these programmes with the representatives of the central and state governments. I recall one such meeting she held with Chief Ministers to review the implementation of programmes for tribal development, on July 13, 1984, where she expressed concern over the lack of involvement by the voluntary organisations in these areas; suggested the involvement of tribal labour cooperatives in place of contractor system; and stressed the desirability of land allotment for the rehabilitation of displaced tribals.

It is, therefore, no exaggeration to say that the greatest contribution of Indira Gandhi in the field of economic planning was to make the whole planning process relevant to the common people of the country, whose expectations and enthusiasm were aroused and whose active involvement in the movement for social justice became real.

She was not oblivious to the ecological degradation caused by reckless denudation of forests and the resulting problems of floods, soil erosion and loss of livelihoods of the poor, especially tribals. Development without destruction became a guiding principle of planning with her active intervention and guidance. Denudation of forests slowed down because of special measures undertaken for afforestation and social forestry. Clearance of development projects from the environmental angle became mandatory and a separate Ministry of Environment was created with necessary institutions to oversee the observance of mandatory provisions.

Such development activities undoubtedly helped to isolate the fundamentalist forces, whether religious or Left-Wing extremism. If the latter has emerged as the biggest threat to national security now, it is for introspection whether these programmes have been pursued with the same vision and commitment as Indira Gandhi had done. It is on the strength of the performance of such programmes that she came back to power in 1980. It appeared then that the whole nation stood as one behind her, the core support coming from the poor and the under-privileged.

As a reaction to the increasing rural-urban and rich-poor divide in the post-reform period, the disenchantment with the 'India Shining' campaign grew a decade ago, culminating in the UPA-I coming to power in 2004 with a promise of achieving 'Inclusive Growth' by pursuing economic reforms with human face in the Eleventh Five year Plan, for the first time after 15 years of economic reforms. This did raise hopes of pro-poor orientation to development with due concern for the protection of environment.

The experience has been encouraging in so far as the growth rate of the economy was stepped up before the on-set of the global financial crisis, and several measures for ensuring inclusiveness such as the Right to Information Act, National Rural Employment Guarantee (NREG) scheme, Forest Rights Act, etc., were introduced. However, it was overlooked that in the prevailing sociopolitical milieu in India, lacking in appropriate regulatory framework and accountability mechanisms, the emphasis on releasing the 'animal spirits of entrepreneurs' when the economy was already on a high growth path could easily degenerate into lending legitimacy to private 'greed' and other predatory practices for amassing wealth by flouting the established norms and regulations. This has been made possible by the increasing collusion of elected representatives with private interests, politicians turning into wealth-seekers and vice-versa. It is good to remember in this context that the need to unleash the 'animal spirits' by stimulating the economy through public investments was first advocated by J.M. Keynes during the Great Depression of the 1930s.

Poverty eradication and conservation of environment which used to be on top of the governance agenda of Indira Gandhi were pushed to the back seat. Poverty level, narrowly defined as those below a certain level of consumer expenditure per head, has no doubt been declining but income inequalities and inequality of opportunity arising from questionable practices have been rising significantly. The socialist or social democratic vision—part of Indira Gandhi's legacy—implies not merely a reduction of poverty defined narrowly, as above, but a social order that is more equitable, just and healthy.

The country is about to launch its Twelfth Five Year Plan with the objective of Faster, Sustainable and More Inclusive Growth. One hopes that the country's leadership and governance would move beyond the usual reiteration of good intentions, and undertake a thorough reappraisal of the underlying social relationships and political economy in the country and come out with a bold and credible package of measures with the necessary political will to arrest the current drift with a view to pursuing economic growth within a just and healthier social order.

Based on the presentation made by the author at the meeting on the occasion of the Twenty Fifth Anniversary of Mrs. Indira Gandhi, organised by the Intellectual Cell of the Andhra Pradesh Congress Committee, in Hyderabad on 30 October, 2009; published in *Hans India*, Hyderabad, October 30, 2012.

30

Rajiv Gandhi

A Pioneer in Economic Reforms

I first met Shri Rajiv Gandhi when I was in the Planning Commission and he, as the General Secretary of the All India Congress Committee (AICC), invited me to deliver the Keynote Address at a seminar on 'Development of Agricultural Sector with Specific Reference to Small Farmers' organised by the AICC on 20 May, 1983, at Vigyan Bhavan, New Delhi. I found him very courteous and warm, and keen on clarifying issues confronting agriculture and the role of small farmers. In his own presentation, among other things, he emphasised the need to extend the Green Revolution to new areas like the Indo-Gangetic Plains and bringing banking to the doors of farmers. Giving the example of his own constituency (Amethi) where recovery of credit was 65 per cent, he thought recovery would not be a problem when credit is productively utilised. He also thought that our education system must be geared to the farmers' needs as, otherwise, they would leave farming when educated.

Immediately after he assumed office as Prime Minister (PM), following general elections, I felt that, perhaps, members of the Planning Commission should submit their resignations, as the new PM may like to reconstitute the Commission. But there was also a different opinion that there was no need to do so, as the new PM belongs to the same party which has been in power. I then hurriedly consulted Shri Nitin Desai who has been Advisor in the Planning Commission for a fairly long period. He was very clear and categorical that propriety demands submission of resignations. He even pointed out that such occasions did arise in the past and members had to resign. Without losing time and in order to make doubly sure, I immediately telephoned to Shri Krishnaswamy Rao Saheb, Cabinet Secretary, and told him that I am thinking of sending my resignation. He immediately said 'you are doing the right thing'. Upon this I told him that my resignation letter would be there within a few minutes. The same day in the evening I received a call from Mr. Alexander, the then Principal Secretary to the PM, saying 'the PM would like to continue you in the Commission' and whether I would be willing, upon which I conveyed my acceptance.

Within a few weeks of taking over as PM, Shri Rajiv Gandhi held an informal meeting with the members of the Planning Commission. He was forthright and appeared very concerned and restless. He said he would like to fully speak out his mind now and members can have their say on a later occasion. What he said then with some passion can be summarised briefly as follows: Planning Commission is basically involved in the routine work of moderating the proposals from Ministries and putting them together as a Plan. As such, Planning Commission does not do planning. Planners ought to be looking ahead at future scenarios for sectors like energy, water, etc., for choosing the efficient path among different alternatives, considering their relative costs and benefits and other relevant issues. The Planning Commission should become a think tank rather than reduced to a bureaucratic machinery.

It was clear from his utterances at this very first encounter with the Planning Commission that he was forward-looking, highly technology-conscious, keen on inducting talent and expertise in the processes of planning and decision-making and moving fast by cutting down delays. His vision was one of managing the economy in a better and more efficient way. In this sense, he anticipated economic reforms of the early 1990s, but with a zeal for technological change, raising productivity and efficiency, decentralisation and peoples' participation. Very soon it became clear that he had high hopes from the introduction of new technologies like information technology (IT) and biotechnology, on the one hand, and from decentralised management by empowering people at the grass roots, on the other. He was deeply suspicious of power brokers and middlemen, as they caused leakages of resources as well as distortions in their allocation.

It was also clear that he was no ideologue, and was pragmatic. For instance, at the first full meeting of the Planning Commission after he took over, held on February 7, 1985, he stated that there was a need for re-look not in terms of basic ideologies but in terms of translating effectively ideologies and policies. There was a need to think big and ahead and to provide outlets for individual initiatives. He was not against subsidies as such. He said food and fertiliser subsidies were a fact of life and so it would not be practical to remove them altogether.

After he took over as PM, there was a move to reopen the question of the size of the Seventh Plan with a view to reducing it, even though it was approved by the National Development Council after considerable deliberations in the full meeting of the Planning Commission, with Mrs Indira Gandhi as Chairperson. But Rajiv Gandhi did not enter into such a debate again. Instead, after checking the size of the two earlier plans, he felt that there was no basis for pessimism and it was difficult to cut plan size, as, according to him, cutting down was no answer; it would amount to an admission of failure and would only build frustration. He was for reducing fiscal deficits but not by cutting Plan size. He said he was against lowering the morale of the people by scaling down the goals.

His vision for achieving social justice was to integrate growth with equity without sacrificing productivity. For example, he was against emphasising 'labour-intensive' methods but favoured employment generation through higher, broad-based growth by reorienting employment in those areas which are efficient and productive.

Of the two major initiatives taken by Rajiv Gandhi, *viz.*, introduction of IT and Panchyati Raj Institutions, the former met with great success whereas the latter is still moving at a snail's pace. When he first talked of IT in the mid-1980s, the skeptics far outnumbered the votaries even in the Yojana Bhavan, though skepticism was confined largely to private gossip. I remember Prof. Yashpal, the Chief consultant in the Planning Commission, quietly dropping into my room one day to sensitise me on the benefits of using IT in the execution of anti-poverty programmes like Integrated Rural Development Programme (IRDP) involving millions of poor households, and explained to me how the whole programme can be monitored right from the Yojana Bhavan. Within less than a quarter of a century, this technology has become a major driving force for Indian economy engulfing millions of enterprises, households and even the 'unskilled' labour throughout the country. This is obviously explained by the strength of the technology itself—its immediate, tangible benefits with low cost, without any adverse effects on health and environment.

Decentralisation of development through Panchayati Raj Institutions, on the other hand, was destined to be a slow process because of resistance from the entrenched interests in our society and polity. At one of the meetings of the Task Force on this constituted by the PM, which included L.C. Jain, Pai Panandikar and B.K. Chandrashekar, finding him and some members quite optimistic about the prospects for decentralisation through *panchayats*, I tried to put in a note of caution saying that empowering the rural poor is not easy because of the stranglehold of vested interests like big land owners, moneylenders and contractors. Shri Mani Shankar Aiyar, who was the convener of this Task Force, also intervened to suggest that initiatives are needed simultaneously to weaken the hold of vested interests, if Panchayati Raj Institutions are to succeed in rural areas. To be sure, thanks to the 73rd and 74th Amendments to the Constitution, following the lead given by Shri Rajiv Gandhi, we now have millions of elected representatives at the grass-roots level who, by themselves, are becoming a major driving force for rural transformation. But political will to empower them through the devolution of adequate functions, finances and functionaries is called for.

A number of initiatives were taken during Rajiv Gandhi's tenure towards liberalisation of the economy and its market orientation by dismantling unnecessary controls and regulations in the sphere of industry and trade. For this, the recommendations of the Economic Administration Reforms Commission (EARC) headed by Shri L.K. Jha, of which I was a member for a year before I joined the Planning Commission in April 1982, were implemented. In fact, it was widely rumoured that the appointment of the Jha Commission by Mrs Gandhi in 1981 was itself prompted by the wishes of Rajiv Gandhi when he was the General Secretary of the AICC. If Rajiv Gandhi was a pioneer among the Prime Ministers in initiating and implementing economic reforms, L.K. Jha was tenaciously so among civil servants.

With the measures for liberalisation of economy getting implemented in quick succession and with hardly any new initiative visible for removal of poverty, the impression was gathering momentum that Indira Gandhi's pro-poor policies were getting a short shrift and the Government was becoming increasingly pro-rich. This was becoming evident from the attitudes, approaches and utterances of the political leaders as well as the senior bureaucrats. Around this time, I was about to complete 5 years in the Government, including a year at the Jha Commission. By this time, I was overtaken by fatigue and a lack of enthusiasm to continue. I decided to resign from the Planning Commission to resume my academic work at the Institute of Economic Growth (IEG).

Four years earlier, I felt a similar lack of enthusiasm, if not fatigue, at the Jha Commission, even before I completed a year there. This was primarily due to my reservations on some of the reforms being recommended which I thought were too radical and did not also feel comfortable with a negative role in which I found myself for most of the time, even though those reforms appear, in retrospect, as too mild and timid. Nevertheless, despite my personal relations with Shri Jha being cordial, I had made up my mind to leave the Commission after the completion of 1 year and get back to the IEG. I had even conveyed this informally to Dr Arjun Sengupta, Secretary to the PM.

A little later, when Prof M.S. Swaminathan left the Planning Commission to take over as the Director-General of the International Rice Research Institute (IRRI) at Manila, I was asked to join the Planning Commission. I joined the Commission with a measure of enthusiasm, as, in the first place, the idea of development planning had been a source of inspiration for all my research work in economics and subjects like agriculture, rural development, welfare of scheduled castes (SCs) and scheduled tribes (STs), backward area development, etc., which I was asked to deal with in the Commission were close to my heart. I was also entrusted with the work on perspective planning. These together with my membership of the Eighth Finance Commission and Economic Advisory Council to the PM kept me fully occupied. All these responsibilities culminating in the finalisation of

Seventh Plan in the second round left me in a state of total exhaustion. In fact, Prof V.M. Dandekar in the course of a meeting with me at the Commission in those days, seeing me tired, sarcastically remarked, "You should leave the Planning Commission before you are fully drained"!

Apart from fatigue and lack of enthusiasm, I could never think of asking the IEG to extend my leave beyond 5 years, having implemented such rules myself strictly when I was Director there, even though Dr Manmohan Singh, the Deputy Chairman, wanted me to continue in the Commission and offered to try for the extension of my leave from the IEG. Thus while my decision to leave the Planning Commission was driven by a number of factors, including, of course, the prospect of pursuing research work using my exposure to policy making in the Government, I was struck by the general tendency to attribute such decisions to a single factor, in my case this being my unhappiness with the new economic policies or differences with the Planning Commission. After sending my letter of resignation to the PM I had to wait for quite some time to get his response, and there were indications from him, then in circulation, that I may have been driven to this step because of some problem within the Commission. This was evident from a telephone call I received from Ms. Sarla Grewal, the Secretary to the PM, enquiring whether I was leaving on my own, to which I said 'yes' and requested for an appointment with the PM.

When I met him he was very cordial and pointedly asked me whether I have any problem in the Commission. At this, I told him that I have no such problem, that I have already spent 5 years in the Government, and now that the Seventh Plan formulation is complete, I am keen to get back to academic life, adding that I would be happy to assist him from outside. He immediately nodded with a gentle smile and began discussing the importance of technological change for raising productivity and efficiency. On my part, I tried to supplement his observations with the need for institutional changes for ensuring peoples' participation and accountability in governance.

Well after I returned to the IEG, one could sense popular discontent in the country with economic policies, as they focused almost exclusively on liberalisation and market orientation. The emerging unrest led to the revival of policy interest on issues of social equity. As part of this renewed interest, Rajiv Gandhi tried to revive Indira Gandhi's 20-Point Programme by appointing an expert Advisory Committee on its implementation, asking me to be its Chairperson. He also constituted the National Commission on Rural Labour.

While his charm and earnestness were disarming, he also occasionally showed impatience, especially when he saw resistance to change from entrenched attitudes. Otherwise he used to be quite informal and jovial in meetings and enjoyed an occasional friendly dig at colleagues, but with a sense of humour. On one occasion in the company of the members of the Planning Commission, he said, "with all respects to the senior economists present here, I would like to mention a saying that economists are those who are heavily paid for making forecasts which often go wrong!". At this, Dr Raja Chelliah who was equally forthright and relaxed gently retorted, "Sir, since we are not paid heavily, our forecasts can never go wrong". The PM greeted the remark with a hearty laugh.

A very genuine and eloquent piece of appreciation for Rajiv Gandhi's personality came from Shri N. Sanjeeva Reddy after he retired as the President of the Republic, when I and my wife, while on a visit to the drought-stricken areas of Anantapur Distrct in March, 1986, paid a courtesy call on him at his native village. He and Mrs Reddy received us warmly and in the course of the conversation, he mentioned about the visit of Shri Rajiv Gandhi to their place only a week ago on his way to Tirupati. Pointing to the place where Rajiv Gandhi sat, he described him as a very sincere young man who had inherited the great qualities of Pandit Jawaharlal Nehru.

The last time I and my wife met Shri Rajiv Gandhi was at an investiture ceremony where Shyam Benegal and Dilip Kumar were awarded Padma Bhushan by the President of India. This was when he had been out of power for some time while Shri Chanrashekhar was the PM. When we saw him standing alone in a relaxed mood, we moved towards him. He immediately recognised us, gave a warm hand shake, and enquired from me about the progress of work on the National Commission on Rural Labour which I was chairing, and whether we are able to come out with useful recommendations. I told him we have made good progress and are busy giving final touches to our recommendations.

31

Economic Reforms with Human Face

A Tribute to P.V. Narasimha Rao

SHRI P.V. NARASIMHA RAO will be remembered, more than any thing else, for successfully piloting economic reforms in the country and equally significantly, though not as well-understood, for his personal commitment to reforms with human face.

Rajiv Gandhi was the first Prime Minister (PM) to see the need for a basic change in economic policy framework being followed in the country since the early 1950s. I could see in the Planning Commission in the mid-1980s how he was restlessly looking for fresh initiatives and for new ways of doing things. But the fate had snatched him away precisely at a time when change and reform were coming at the forefront of national agenda, not the least, because of compulsions of development elsewhere in the world, especially in the former socialist countries.

This historic task fell on the shoulders of Shri P.V. Narasimha Rao. The unprecedented economic crisis faced by India in 1991, in the context of the then evolving international milieu, left us with no option but to embark upon economic reforms. It is difficult to say whether P.V. garu had anticipated these events or whether he would have undertaken such bold initiatives in the normal circumstances. But there is no doubt that he saw clearly the writing on the wall and responded to the challenge of crisis with courage and equanimity and with a surprising measure of decisiveness which was not a characteristic of him. For this task, he chose Dr Manmohan Singh, an eminent economist with a vast experience in the Government, as his Finance Minister. Dr Singh responded to this challenge with the best of his abilities and successfully coped with the crisis. Together, they brought about a paradigm shift and systemic changes in the economy, unprecedented in their consequences since the days of Pandit Jawaharlal Nehru.

I had known P.V. garu right from my student days in the early 1950s, but I could interact with him closely only much later in Delhi. In the brief interregnum after the assassination of Shrimati Indira Gandhi and until Shri Rajiv Gandhi took over as the PM again after the general elections, P.V. was the Deputy Chairman of the Planning Commission, and I had opportunities of interacting with him as a member. Only a few months earlier, Shrimati Indira Gandhi had defined the objectives of the Seventh Plan as 'Food, Work and Productivity'. P.V. was keen that we should give employment orientation to our plan programmes. I remember discussing with various divisions in the Planning Commission to explore the possibilities of giving employment orientation to the Seventh Plan. Because he had earlier initiated bold measures for land reform as the Chief Minister of Andhra Pradesh, his credentials as one who would work for social justice were strong. However, despite such credentials, I was not too sure that while undertaking economic reforms in the early 1990s, the concerns for employment generation and social equity were being adequately addressed. I did express my misgivings on a number of occasions to both PM P.V. garu and Finance Minister Dr Manmohan Singh. P.V. garu in the course of one such discussions tried to assure me that these concerns are being taken care of. He said that he was quite alive to this problem and that was precisely the reason why he brought into the PM's office distinguished and committed officers like B.N. Yugandhar and K.R. Venugopal. His sincerity on this count is beyond question, as he realised the importance of rural development and was the first PM to keep this portfolio with him and also allocate a record amount of resources for rural development, when Shri S.R. Sankaran, was the secretary in this Ministry.

The Scheme for Employment Assurance in the backward districts was introduced by him. He explicitly pleaded for the 'Middle Path' in economic reforms at the international economic summit in Davos. He successfully piloted 73rd and 74th Amendments to the Constitution, initiated by Rajiv Gandhi, for empowering the local elected institutions. He promptly accepted and implemented the recommendations of the Technical Committee (with which I was associated as chairman) for conservation of land and water resources through participatory development of watersheds. Despite all this, the impression persisted that while pursuing economic reforms, the concerns for social justice were not being adequately addressed, probably because the measures for globalisation and liberalisation were prompt and well-articulated whereas those for social justice were halting and muted.

Whether it was on account of wisdom born out of experience or because of freedom from the constraints of a 'reform-bound' Prime Ministership, P.V. garu became quite critical and was unusually articulate about the reform process since the late 1990s. He spoke against the reckless drive for privatisation of public enterprises and argued for greater concern for reducing regional disparities and for social justice in general. At a seminar, he organised on Regional Disparities in Development under the joint auspices of National Institute on Rural Development (NIRD) and the Swami Ramanand Teerth Research Institute, which he headed, he invited me to speak on the subject as a panelist along with late Shri Govinddas Shroff, Shri P.S. Krishnan and others, where I found him very meticulously discussing measures for reducing regional disparities in development.

The last time I met P.V. garu in Delhi was about a year before his passing away when I presented to him a copy of the book on Andhra Pradesh Development brought out by the Centre for Economic and Social Studies, Hyderabad. He appreciated the effort and showed his eagerness to read the book. The very next day I learnt from his long-time colleague Shri Keshavulu garu, Director of Swami Ramanand Teerth Institute, that he received a telephone call from P.V. garu to say that the book is good and that there should be a wider discussion on the issues raised.

The best tribute that we can pay to Shri P.V. Narasimha Rao garu is to carry forward the reform process he had initiated, with a clear and strong orientation to social justice—with accent on employment generation, rural development and reducing regional disparities.

Jayanthi, April-September, 2005, Viswanatha Sahitya Peetham, Sister Nivedita Foundation, Hyderabad.

32

J. Chokka Rao

Grassroots Leader and Policy Ideologue

I had known Shri J. Chokka Rao garu closely for nearly three decades. We used to meet frequently in Delhi since the early 1970s after he became the Chairman of the Telengana Regional Committee. We used to discuss then the measures for the development of the Telengana region and how to present the case effectively to the Prime Minister and to others in the Planning Commission. Our contact got revived again when I joined the Planning Commission in the 1980s and this occasional exchange of views continued until his passing away. He became more vigorous in his consultations with experts when he became Chairman of the Standing Committee of Parliament on Agriculture in the early 1990s. In a sense, this was the most important public position in his political career, which he shouldered with confidence and presented a series of reports on major issues facing agriculture.

I had also known his close associate Shri V.B. Raju when I worked with him for some time in the mid-1950s as a student activist in the movement for the liquidation of the then Hyderabad state and the formation of Vishalandra. I am recollecting these events because one cannot understand and appreciate fully the contributions of Chokka Rao without referring to his acquaintance with V.B. Raju.

V.B. Raju was highly intellectually oriented. He could discuss with sophistication diverse theories, policies and programmes. Chokka Rao, on the other hand, was a grassroots politician without much of formal educational qualifications. Yet, his horse-sense for the real issues of development was compelling, and therefore, seems to have been influenced by V.B. Raju in the pursuit of such interests at the practical policy level. It is thus amazing how these two individuals came so close to each other, and even more amazing, how Chokka Rao took on the role of a policy ideologue in the mould of V.B. Raju.

Being an experienced and much accomplished grassroots leader, he could have easily devoted his life to the more enticing ways of getting the development projects for his constituency and having them executed. But Chokka Rao's sights were clearly higher than the mere pursuit of such activities at the district level. Moreover, he had already done pioneering work for the development of the district as a long-standing leader and as a Chairman of Zilla Parishad. Building on this, his interests became broader.

That is why, when he contested for the Lok Sabha seat for the second time in 1990, he was confident of going to the people with his party's programme. I remember his dropping in at my office in Delhi on the eve of this election when I was Chairman of the National Commission on Rural Labour, and discussing with me seriously the contents of a people-centred programme which he would like to place before the people. I came to know from other friends that he had similar discussions with them. It is indeed very rare these days that a candidate is keen to project a policy and programme before the people rather than giving the doles.

I could see that his interest in economic policy matters was deep and genuine. He was persistent and did not give up this interest. He was conscious of his limitations and had the humility to seek assistance from his friends and colleagues and had the tenacity and confidence to pursue them. He used to work very hard by discussing diverse issues in finance, agriculture and rural development with high-level experts in the country and by checking and cross-checking ideas and material with them. Because of his sincerity of purpose and friendly manners, he could get willing co-operation and help from such experts and intellectuals.

He was a Member of Legislative Assembly (MLA) for several terms; a Cabinet Minister in the State government; and was elected twice to the Lok Sabha. If people showed so much consideration and regard for him, it was because of his image as one who did not make a fortune for himself by using his power and influence; was deeply committed to development and social justice; and was easily accessible.

His articulation—vocal quite often—may not have always gone quite well with some sections of the intelligentsia. It, nevertheless, lent simplicity and transparency to his personality. The story of Chokka Rao highlights the simple truth that, in the ultimate analysis, people like honest individuals with social concerns.

J. Chokka Rao, Commemorative Souvenir, Edited by K.R. Venugopal, May, 2000, Hyderabad.

33

N.T. Rama Rao

From Film Hero to Mass Leader

I had the privilege of meeting Sri N.T. Rama Rao (NTR) garu for the first time in the early 1980s soon after he took over as the Chief Minister of Andhra Pradesh, when he came to Delhi along with his colleagues for discussions with the Planning Commission (where I was a member), to finalise the Annual Plan of the state. He invited some of us for an informal get-together over the breakfast at the Andhra Pradesh Guest House. In the course of these discussions, one could see his concern over the difficult resource position of the state in the face of fresh commitments including his desire to undertake a number of welfare programmes. Some time later, I had an opportunity of meeting him again when the Andhra Association of Delhi decided to honour some of the Telugus in Delhi, at the hands of the President of India, Sri Gyani Zail Singh, in the presence of the Chief Minister Sri N.T. Rama Rao.

NTR prominently strode on the political scene in Andhra Pradesh in the early 1980s when there was considerable political instability and uncertainty in the state. The greatest source of his popularity then was his charisma as a hero in films, especially the mythological ones. True to the image of these roles, people found in him a saviour and a solution to the prevailing instability. But NTR himself did not bank on this particular 'image' for long. His commitment to work for the welfare of the people was strong enough to stand him in good stead, in his own right, as a mass leader.

Soon after assuming office he came out with a series of welfare schemes, the most prominent among them being the supply of rice at ₹ 2 per kg, through the public distribution system (PDS). This scheme has often been questioned for being 'populist' and for the large amount of subsidy it involved. But the merits of the scheme have been well recognised, over time, especially its potential for improving the nutrition levels of women and children among the poorer households. Earlier, M.G. Ramachandran (MGR) had introduced such a scheme in Tamil Nadu which resulted in providing much needed relief to the poor and empowering women managing the households. The Food Security Bill currently before the Parliament involves subsidy on the sale of food grains on a significant scale underscoring the need for the continuance of such schemes in varying forms even decades after they were first introduced.

The second major initiative of NTR was towards empowerment of women through various measures including, particularly, equal right to property and introduction of prohibition. The emergence of women power in the state was conspicuous during his tenure. This led to the organisation of women's self-help groups in later stages which has been applauded as a unique phenomenon in Andhra Pradesh. The steep fall in the birth rate noticed in the state from the Census of 2001, despite lower female literacy and high infant mortality, has been attributed by experts to the emergence of women's power. Earlier, Tamil Nadu and Kerala had experienced a similar decline in birth rate, mainly on account of higher female literacy and lower infant mortality.

The third major concern of NTR was decentralisation of development by taking the programmes to the door-step of the villagers for involving them as participants in the development process. The creation of a large number of revenue *mandals* with elected representatives and the abolition of intermediaries was a step in this direction.

The poor and the underprivileged, especially women, rallied around him in a big way in the state, as he carried immense credibility with the common people. He did not live to see globalisation and liberalisation of the Indian economy in full measure in the 1990s. There is no doubt, however, that he would have been open to these changes. It is equally clear that, in view of his strong commitment to improve the livelihoods of the poor and the under-privileged, he would have tried his best to combine economic reforms with measures for social equity from the very beginning. It took more than a decade for his successors in the state to learn the simple lesson that in an unequal society like ours, the benefits of economic growth may not trickle down to the poor unless bold measures are undertaken for inclusive growth simultaneously while pursuing economic reforms.

Souvenir, NTR Memorial Trust, Hyderabad, April 2013.

34

Abid Hussain

Reform-Oriented and Warm-Hearted

THE last time I met Dr Abid Hussain was on 29th February, 2012, when he delivered the Quli Qutub Shah Extension Lecture on 'Spiritualism, Secularism and Composite Culture of India with reference to Urdu', at the Salarjung Museum, Nayapul, under the auspices of the Department of Urdu, University of Hyderabad. He spoke in chaste Urdu for full one hour without a pause even for a few seconds. I was astonished at this and became diffident as I was to speak after him in Urdu. I did say there that after listening to Abid Saheb's non-stop presentation in chaste Urdu for an hour, I did not have the courage to speak in Urdu.

Having left Hyderabad in 1957 to live and work in Delhi for over three decades, doing all my professional work in economics and economic policy through the medium of English, I had almost forgotten my Urdu and Telugu, which I am now trying to revive in Hyderabad by reading newspapers in Urdu and Telugu. Like me, Abid Saheb too was in Delhi for much of this period using English all the time for his professional communication and so I wondered how he could retain such freshness in his Urdu.

As someone in this audience said in his note circulated, Abid Saheb was a '*pakka*' Hyderabadi. I think, he represented the best in *Hyderabadi Tahzeeb*. While he was speaking on secularism and composite culture of India in Urdu with a natural zest by quoting memorable exchanges between Gandhi and Tagore, I thought Abid Saheb himself embodied fully the values of such a composite culture. He distinguished himself as an Indian Ambassador to the USA and I would like to add that he was also an Ambassador of *Hyderabadi* culture for the rest of India and the world at large. He was warm hearted, very friendly and likeable, witty, encouraged youngsters and helpful to the people around him. All this was natural to him and came from him spontaneously.

I had known Abid Saheb over a long period and had the privilege of closely working with him as a colleague in the Planning Commission in the mid-1980s when the Seventh Five Year Plan was being formulated. I found him reform-oriented and much before economic reforms were introduced in India in the early 1990s, he was for new and better ways of doing things.

Though holding a definite pro-reform stance, he was a moderating influence whenever the controversies reached a high pitch in the course of discussions on plan formulation. He was known for his pro-reform orientation way back in the 1970s and 1980s, as was evident from the recommendations made by a number of important committees on industry and trade, including the one on small-scale industries, under his Chairmanship. He was, of course, for reforms with a human face and social equity. It could not be otherwise for a person like Abid Saheb.

I think, this trait in him dates back to the 1950s when he was an in-charge of the Community Development Programme in Hyderabad. This programme aimed at innovative ways of involving communities for rural development and welfare. I recall how he won over a top communist leader like Shri Ravi Narayan Redddy garu to the idea of Community Development by showing him various activities he was organising. Shri Ravi Narayan Reddy garu himself told me that he was highly impressed by the work Abid Saheb was doing.

Observations made at a meeting held in memory of Dr Abid Hussain, at the Administrative Staff College of India, Hyderabad, on July 8, 2012.

35

Mani Shankar Aiyar

Empowering Panchayati Raj Institutions

SHRI MANI SHANKAR AIYAR was introduced to me by Abid Saheb in the Planning Commission. My association with Shri Aiyar became closer and stronger when Shri Rajiv Gandhi, with Shri Aiyar's active involvement and help, took a major initiative to revive Panchayati Raj Institutions in the country by statutorily empowering them through the necessary amendments to the Constitution. Shri Aiyar organised a task force to work on this and I had the privilege of being invited to join this group. Shri Aiyar, as a convenor, organised a series of discussions of this task force with the Prime Minister Shri Rajiv Gandhi which culminated in the formulation of the strategy and 73rd and 74th Amendments to the Constitution for empowering the Panchayati Raj Institutions.

Mani Shankar Aiyar was known to be close to Rajiv Gandhi since their Cambridge days. It used to be said that Rajiv Gandhi actively supported Mani Shakar Aiyar when he was elected as the President of the Students' Union at Cambridge. After completing his education, he joined the Indian Foreign Service where he had a long experience as a diplomat. A person with this background in foreign affairs getting actively interested in issues of governance at the grassroots level in the country is very rare indeed. As it is, ironically, globalisation turned to be the easiest of all economic reform measures in India whereas decentralisation of governance the most difficult one.

Shri Aiyar combines his brilliance and oratorical skills with his deep commitment to the cause of strengthening democracy at the grass-roots level. His writings on the subject are bold, incisive and compelling. It is only appropriate, therefore, that he should have been chosen as the First Union Minister for Panchayati Raj. As a minister, he took major initiatives to strengthen the Panchayati Raj Institutions in the country, among others, by holding a series of regional conferences of the concerned ministers and officials of the state governments. Given our social structure, particularly in the countryside, and the political forces it has unleashed at the state level, this has been an extremely challenging task, as it virtually involves swimming against the current. One cannot think of a more committed, courageous and astute person than Shri Mani Shankar Aiyar to handle this difficult task.

For some time, Shri Aiyar was incharge of the All India Congress Committee (AICC) refresher Training Programme for the Congress leaders. Shri Aiyar invited me to address some of these sessions. In one such session, I had a pleasant interaction with Sunil Dutt, Member of Parliament (MP), on the issue of rural credit. He was surprised to learn from my presentation that credit advanced by rural moneylenders at high interest rates continues to be substantial. He recalled in this context the picture, 'Mother India', in which he had acted as a young man resisting the exploitation by the moneylender.

While writing about Mani Shankar Aiyar, I cannot resist the temptation of recalling my acquaintance with his brother Swaminathan S. Aiyar, one of the most well-informed and effective economic writers in the country. In fact, thanks to our common interest in economic policy issues, I have had opportunities to interact with him over a longer period than with the elder Aiyar. Ideology-wise, these Aiyar brothers appear as poles apart, though equally effective in their respective spheres. But, those interested in pursuing 'reforms with human face' would be looking for a policy framework which reconciles or integrates Mani's concerns for 'equity' with Swami's prescriptions for 'efficiency'.

36

B.P.R. Vithal

Grip on State Finances and Founding of the CESS

PROF V.K.R.V. RAO once mentioned to me about Sri B.P.R. Vithal garu very warmly as one who chose to work for the state government with commitment disregarding enticements for working at the Centre. Sri K. Rosaiah garu, the veteran Finance Minister in the state—now Governor of Tamil Nadu—told me that Vithal garu, because of his expertise in planning and finance and the confidence he evoked, successive governments of diverse political persuasions found comfortable working with him. Apart from his grip on the subject, this speaks of his ability to convince the leaders on the merits of the issues at hand as well as accommodate the concerns of the elected representatives without compromising on the reasonableness of the proposals or the soundness of allocations.

The way he was trusted by the successive Chief Ministers and Finance Ministers enabled him to create an autonomous research institution like the Centre for Economic and Social Studies (CESS), initially with full financial support from the state government. The main objective of this Centre at that time was to do all the technical work needed for the Planning Department. Happily, a good deal of such work continues to be done at the Centre, to this day, thanks to the active interest shown by the secretaries of Planning as well as Finance who are also the members of the Board of Governors of the CESS.

But Vithal garu's perspective was much wider. As he once explained, he was also inspired by the vision of Professors V.K.R.V. Rao, K.N. Raj and Malcolm Adiseshiah to decentralise research away from Delhi for focusing on the problems at the grassroots level. Our experience has been encouraging as the themes chosen for research at the CESS are relevant to the region, such as agriculture, poverty and environment, and there is a good interface with scholars working in other research centres and civil society organisations in the state.

Over the last two decades, including 12 years when I was the Chairman of the Board, I have had several occasions to closely interact and work with Sri Vithal garu. I am struck by the active interest he has been taking, to this day, in the development of the Centre and, I must add, he has been doing this with a rare sense of detachment from its day-to-day functioning.

The sources of inspiration for his social commitments and widening of his horizons are deeper. Not many in this audience may be knowing that when he was barely 15, he chose Mahatma Gandhi's teachings as a way for his life style, corresponded with him, did spinning regularly, wore *khadi*, attended the Quit India Session of the Congress at Bombay in August 1942 and was even witness to the number of votes cast for and against the Quit India resolution. His political digressions caused a 3-year break in his education, but, according to him, these have been both exhilarating and elevating.

This is a day for our thanks giving to Sri B.P.R. Vithal garu and for resolving to carry forward his legacy.

Observations made at the *National Seminar on Centre–State Relations in the Fiscal Context* (In Honour of Sri P.P.R. Vithal, Founder, CESS), Centre for Economic and Social Studies, Hyderabad, November 5-6, 2012.

37

Y.V. Krishna Rao

A Rare Peasant Leader with Scholarship

I have great pleasure in extending, heartily and respectfully, my felicitations to Com. Y.V. Krishna Rao garu on the occasion of his 87th birthday. My personal acquaintance with him started nearly 25 years ago when I was working at the Institute of Economic Growth (IEG), Delhi. This association continued when I moved to the Planning Commission and after my return to the IEG. My association became even closer for nearly 10 years now after I settled down in Hyderabad. During all these years, I had several occasions to exchange views with him on many issues facing Indian agriculture and the role of the peasant movement in the context of the changing rural scenario.

What struck me most about him is his passion for learning and his burning desire to be up to date on the vital issues facing Indian agriculture. His passion to be in touch with the latest material on the subject was not motivated by the narrow instinct to find arguments and data for substantiating any preconceived or settled arguments as is very often the case with many leaders. His genuine desire to understand reality and subject his own ideas to scrutiny, where necessary, was transparent in his pursuit of the latest information and thinking on the subject.

He combined his quest for being in touch with emerging facts and trends with his deep understanding of Indian agriculture and his vast experience of peasant movement in the country. This is a rare combination and, so, has lent authenticity and credibility to his voice in seminars and conferences. He has always been alive to changing situations and appeared ready to look at the events with an open mind.

His genuine quest for knowledge, his humility and genial disposition brought him closer to several leading agricultural economists in the country and appeared quite comfortable in their company in serious academic seminars. He was also a member of the Commission on Agricultural Costs and Prices (CACP). He displays all the attributes of a professional scholar and is undoubtedly one among them. He is extremely hard working and in this respect, can put to shame many scholars who are decades younger to him.

Sri Y.V. Krishna Rao 87th Birthday Felicitation Volume, 15-09-2001, Hyderabad (Edited by Sri N. Shiva Rami Redy and Dr S.V. Satyanarayana)

38

Dr Raj Bahadur Gour

A Veteran Trade Union Leader

DR RAJ BAHADUR GOUR has behind him half a century of experience as a freedom fighter and trade union leader. I and several other students of my generation in the mid-1940s were greatly inspired by his sacrifice of a lucrative medical practice for the sake of participating in the freedom movement, and not the least, by his eloquence in Urdu laced with devastating wit.

After 50 years of workers' struggles and changes in the economy and polity, I guess, the large majority of workers under the organised trade unions that he is leading now must be above the poverty line. A large majority of the working people are still poor and subject to various forms of exploitation. Only that they are not organised and fall into the category of informal sector, both urban and rural, where contract labour is significant without any social security benefits. The trade union movement can take a great deal of credit for the improvement in wages and living conditions of the organised working class. But much of this improvement owes also to the changes in the economy and polity.

Firstly, economic development and technological changes in general have contributed to the rise in productivity and surpluses of the enterprises facilitating the sharing of these gains with the working class, albeit, in response to the struggles waged by them. Secondly, the almost uninterrupted working of the democratic system ever since independence has ensured to our people at large freedom and democratic rights for resisting exploitation and injustices. Finally—and this is equally important—the trade union leadership itself has increasingly moved to the sectors where bargaining power exists and hence prosperity is more easily won. Some occupations are highly skilled and strategic. Being strategic, the workers there enjoy greater bargaining power. Being highly skilled, their wages have necessarily to be high. With economic development and modernisation, the importance of such sectors has increased over a period of time.

What are the implications of the emerging economic scenario, particularly in India, for the trade union movement?

There is no conflict between high wages and achievement of high levels of development. In fact highest wages and shortest working hours are found in the most developed countries. However, there are certain 'disquieting' implications for the trade union movement. The recent technological changes are making small-scale decentralised activities economically viable. More significantly, the importance of the service sector in relation to manufacturing has been rising phenomenally in the course of economic development. This has resulted in the fragmentation of labour force with growing number of part-time workers and flexible working hours. Even in the developing countries, there is a shift of workforce away from production line towards more fragmented units in the services sector. Capital is also moving from countries with powerful trade unions to countries where wages are low and workers are not sufficiently unionised.

As a consequence of all this, the trade union membership as a proportion to total number of workers seems to have fallen significantly in the developed countries of the West. In the developing countries like India, relatively smaller proportion of workforce is unionised because of fewer workers in the formal sector. The current drive towards privatisation in the wake of globalisation threatens to weaken the trade union movement. Already, since the mid-1980s, the bargaining power and the 'militancy' of the employers seems to have increased more than that of workers. This is indicated by the steep rise in the man-days lost due to lockouts in relation to those lost on account of strikes.

In India, the majority of workers are outside the trade union movement. They are much poorer and severely exploited. Trade union movement has always been an integral part of the overall movement for ensuring the well-being of working people in general in the country. If this objective is to be pursued seriously, the trade union movement has to concern itself with greater employment generation in the economy, on the one hand, and improvement in the living conditions of the unorganised sections, on the other. The first objective requires building up political pressures for an appropriate policy framework and the second objective requires direct involvement of organised workers and their leaders in educating and organising the unorganised sections of workers.

This should include developing awareness about public programmes for upliftment of workers and active assistance to enable them to derive benefits from such programmes. Equally important would be the possible contribution of organised workers in vocations which directly affect the living conditions of the poor. For example, the workers' unions in the commercial banks can try to ensure that the poorer household enterprises have adequate access to credit. In rural areas, for instance, bulk of the credit from such banks is cornered by the affluent sections at concessional rates of interest leaving the poorer households to borrow from the informal credit market at exorbitant rates of interest. The unions can educate their members in the banks to ensure the provision of such services on an equitable basis.

Excerpts from the paper presented at a seminar on the occasion of the 75th birthday of the veteran freedom fighter and trade union leader, Dr Raj Bahadur Gour, held at the Urdu Hall, Himayat Nagar, Hyderabad, on July 22, 1993.

39

Professor K. Jayashankar

A Leading Telangana Ideologue

WITH the passing away of Professor K. Jayashankar, the movement for the formation of Telangana state has lost the foremost academic and intellectual, actively working for this cause for over half a century. The void created by his demise cannot be filled except by imparting to the whole movement the values he embodied: a firm commitment to the cause of separate statehood combined with a highly informed and reasoned approach to the many issues faced by the movement.

In many ways, Prof Jayashankar revealed a Gandhian approach in his working. From the very beginning, starting from the Mulki agitation of 1952, he had a steadfast belief in the justness of the cause he was espousing. He was always laborious and careful in marshalling the relevant data and intellectual reasoning in support of the cause. Like Gandhi, he was fearless and fully confident of facing and convincing his adversaries on the relevant issues. This made him travel throughout Seemandhra several times to discuss, argue and try to convince the intellectuals on many issues concerning the need for statehood for Telangana; on how this will not adversely affect the interests of Seemandhra; and on the contrary, how both the Telugu speaking states can develop faster in an environment of harmony. This mission also took him abroad, particularly to several places in the USA. He also played an active role in Delhi towards convincing and persuading the leaders of various political parties to send their letters of consent for the formation of Telangana state, to the Committee headed by Pranab Mukherjee. Again, like Gandhi, his style was one of relentless persuasion through data and arguments without any trace of bitterness. Finally, he was totally selfless and free from any desire for position or power. His gentleness and humility were disarming. All these qualities made him a strong exponent, endearing and credible.

This indeed explains the unprecedented and overwhelmingly touching tributes paid to him by the people of Telangana.

In view of his life-long academic career, culminating as the Vice-Chancellor, Kakatiya University, Warangal, he would perhaps have preferred to keep himself away from active politics. But given his strong commitment to the formation of Telangana state, he had the practical wisdom to see that the pursuit of such a goal requires not merely intellectual advocacy through writing papers and books, but a meaningful participation in a promising movement towards achieving this goal by providing intellectual inputs. The subsequent developments have fully vindicated the wisdom of his decision. The Telangana movement today is stronger than ever before, not the least because he and some of his academic colleagues have thrown their lot into this movement. It is indeed tragic that he should have been taken away precisely at a time when the Telangana movement is at its peak, when his advice and leadership are needed most and when there is every likelihood of himself seeing his dream fulfilled.

Mainstream XLIX (29), July 9, 2011.

40

Baddam Yella Reddy

A Crusader against Feudalism

IT was 1943-44 amidst the Second World War. I was hardly 14-15 years old, staying with my sister and brother-in-law (B-in-L) in Karimnagar, studying in the nearby school adjacent to the imposing clock tower which used to ring every hour resoundingly making us time conscious. My B-in-L Sri Polkampalli Venkata Rama Rao garu was a leading and powerful lawyer. He was a staunch nationalist and his office room was adorned with large size portraits of leaders like Mahatma Gandhi, Jawaharlal Nehru, Motilal Nehru, Lala Lajpat Rai, Chittaranjan Das and others. He also became president of Andhra Mahasabha, then dominated by the leaders of the growing communist movement in Telangana.

It was here that Sri Baddam Yella Reddy garu (BY) used to drop in almost every evening for political gossip. He used to be accompanied often by Sri A. Prabhakar Rao garu—a very sincere, militant and outspoken communist leader and a martyr in the Telangana Armed Struggle—for whom my B-in-L had a great liking. I used to be around and be a mute witness to these discussions, partly because I would be called frequently by my B-in-L to get for him from the bazaar Charminar Cigarattes—perhaps an essential ingredient for warming up the discussions!

These discussions, I vaguely recall, used to be about the on-going World War, the heroic role of the Red Army, about the communist party of India's stand towards the Congress in the freedom movement and on several issues of national importance.

It was the impact of such intermittent sessions that may have brought me under the influence of BY and the communist ideology. This was ironical because I was staying with my B-in-L, who was also my guardian, and there was a liberal, intellectual atmosphere at home with very distinguished visitors from different parts of the country dropping in. In retrospect, I can recall and connect a series of factors leading to such influences overtaking me. This could also be true of my elder brother Sri Ch. Rajeshwara Rao garu who was around studying in Karimnagar and visiting B-in-L quite frequently, the latter having a strong liking for my brother.

The simple and mingling behaviour of communist leaders like BY with the youngsters could be one such factor. But the 'pull' factors were many. The 'nationalist' call was basically for achieving freedom from foreign rule. The communist movement too was championing this cause with equal force but promised much more, especially the elimination of feudalism and formation of a socialist society. This had a great appeal those days, especially in Telangana which was reeling under feudal oppression.

This may well explain why the communist movement was much stronger than that led by the Congress in Telangana before the police action against the Nizam by the Government of India. Apart from this broader vision, the communist movement had a consistent and coherent world outlook, though later this view proved to be too simplistic. All the same, the sensitive and brighter sections of youth those days were attracted more by the wider world outlook than the one limited to one's own region or the country. In Karimnagar, at any rate, another reason clearly inducing me towards the communist movement was the flood of literature from the Prajashakti publications. The Congress stream, on the other hand, had very little to offer locally that was readable immediately, apart from the news of an aura surrounding the nationalist movement and its leaders outside Hyderabad state. There was a '*Krishna Patrika*' locally available and an exhaustive daily like *The Hindu*, which the youngsters like me were ill-equipped to read and understand at that time.

It was under these circumstances that BY emerged as one of the topmost leaders of the Telangana movement and undoubtedly the tallest figure in Karimnagar district. He was the first *satyagrahi* in the district and was imprisoned or intimidated by the police several times. The youth was inspired then more by the leaders who were dedicated to the great cause and made sacrifices than by those known mainly for their intellectual calibre or oratorical skills. The stories of how BY used to treat the police as dirt and full of scorn whenever they approached him for an enquiry or for taking him into custody were popular and remembered with great relish. The mass appeal of the movement he led was at its peak at the time of the first general elections in 1952 when he was elected to the Lok Sabha from Karimnagar under the auspices of the Peoples' Democratic Front (PDF) with a large majority over the Congress candidate, Shri P.V. Narasimha Rao.

Like many freedom fighters of that era courting jail intermittently or going underground, BY could not acquire high academic qualifications. But, my close interactions with him proved, beyond doubt, his intellectual capabilities, whether it is sharpness in reasoning or organising one's material systematically or consistency of presentation. One of the very first set of books in Telugu I read in 1945 or so was his book entitled '*Neti Hyderabadu—Mana Kartavyam*', meaning 'Hyderabad Today—Our Tasks', co-authored by his colleague Sri Devulapalli Venkateshwara Rao garu. This was considered as the classic work on the Hyderabad state and was invariably used in the training courses (see below my comment on the book). I myself attended one such course in Karimnagar where BY himself took several classes. He was known to be very meticulous in looking at the drafts given to him and it was very difficult to get them cleared by him without satisfying him on many queries that he would certainly raise. He was a voracious reader and spent considerable time for this every day.

I was constantly in touch with him as an activist of the All Hyderabad Students' Union and had the benefit of his advice and guidance right from the Karimnagar days in 1945-46 until I left Hyderabad for Delhi in 1957 for doing my Ph.D. He used to closely follow the activities in the student movement and give his assessment and suggestions. He would often get impatient with injustices done to the students and would urge for some immediate action by the Students' Union, but would understand the constraints when properly explained.

When I decided to leave for Delhi to continue higher studies, which really meant a break in my career from politics to academic profession, BY did not show any unhappiness on my 'leaving' the movement and, instead, saw something good happening. He encouraged me to pursue higher studies. He was fully aware of the growth of various academic disciplines and their contribution to the society. Thereafter, we used to meet occasionally either in Hyderabad or Delhi and engage in discussions and arguments on India's agricultural economy. Once in 1963, he came all the way from New Delhi to our one-room apartment on the third floor in Kamlanagar when I and my wife had the privilege of having lunch with him. This showed how he was fond of me. Gradually I lost contact with him.

But, I was deeply touched when Vijaya, BY's daughter, and her husband Narahari Rao met me the other day to extend invitation to attend the marriage of Baddam Lakshman Reddy's son. It was a sort of homecoming for me. The last time I used to see her brothers Ram and Lakshman was in the early 1950s, when they were just a couple of years old, at the old Member of Legislative Assembly (MLA) quarters with their parents, where I was staying as a neighbour with my maternal uncle Sri J. Ananda Rao garu, who was elected as an MLA in the first general elections from the Sircilla constituency. On this occasion of the birth centenary of Shri Baddam Yella Reddy garu, I wish the Baddam family all the best.

Baddam Yella Reddy 1906-2006, *Birth Centenary Volume (Telugu)*, Baddam Vijaya (Ed.), Hyderabad, 2006.

41

Hyderabad Today: Our Tasks

A Comment

It was an exciting experience for me to read this precious book in Telugu again after 60 years, both because of my familiarity with the then sociopolitical scene in Hyderabad and my close acquaintance with the coauthor of this book, Shri Baddam Yella Reddy garu. (The other author is Shri Devulapalli Venkateswara Rao.) The book portrays vividly the highly exploitative feudal land tenure system, the despotic rule of the Nizam and the emerging peoples' movement spearheaded by the Andhra Maha Sabha in Telangana against feudalism for the formation of a responsible government and for tackling the problems of hunger and poverty.

The book has a broader perspective, relating the political developments in the Hyderabad state with the world-wide anti-fascist struggle during the Second World War and the national movement for independence. The food problem in the state is especially highlighted as the warning from the dreadful Bengal Famine of 1943 was loud and clear.

The book also emphasises the imperative need for Hindu-Muslim unity and the broadest possible peoples' movement against fascism and for solving the food problem.

On the agrarian front, the book extensively documents how the peasantry was burdened with high taxation of land and innumerable taxes on professions; high rents under insecure land tenures; very low wages for agricultural labour; non-availability of credit and exorbitant interest rates charged by the moneylenders; and persistence of primitive methods of cultivation as well as degradation of land due to illiteracy and poverty.

The book and the movement it mirrored turned out to be prophetic in several respects. With the successful conclusion of the Second

World War following the defeat of fascism, the political situation in the country as well as in the state changed dramatically. Within 2 years after this book was published, the anti-feudal movement turned into a radical campaign for 'land to the tiller' through armed struggle in Telangana. Barely 3 years after this book was written, the country achieved independence and, a year later, the police action against the Nizam resulted in the merger of the Hyderabad state with the Indian Union. These historic developments showed the many dreams in the book coming true in one go—perhaps much earlier than even the leaders of the movement anticipated.

Despite these great accomplishments, the message of the book continues to be relevant even to this day because of the unfinished rural agenda• on which there is a broad measure of consensus in the country: the plight of the small and marginal farmers with widespread tenancy, high rents and rising risk; inadequate access to rural credit and high interest rates; slow improvement in methods of cultivation and land degradation; persistence of rural poverty and malnutrition; and weak local self-government due to tardy devolution. The solutions suggested in this book are as valid today as they were 60 years ago.

42

My Nizam College Days (1953-1955)

I CHOSE NIZAM COLLEGE for pursuing my graduate studies in Arts in 1953 solely because of its reputation as a premier institution with English as its medium of instruction. Luckily, I could get admission having passed intermediate as a private candidate from the Board of Secondary Studies, Ajmer. This was after 4 years of interruption in my studies following my arrest in February 1949 for my participation in the student movement, when I was in intermediate (Science) final year at the City College.

At Nizam College, I found myself in the company of bright students—much younger than me—speaking English fluently with impressively high level of general knowledge, for their age. It became clear to me soon that the reputation earned by the Nizam College was no less due to the quality of students, many of whom had the good fortune of going to first rate schools in the city.

The cosmopolitan character of student community there was glaring. Hyderabad being the capital of a multi-lingual state, we had a large number of students from Marathwada and Karnataka, apart from Telangana. Also, we had a good number of students from the North speaking Hindi and from the South speaking Tamil and Malayalam. Equally striking was the diversity of their religious composition.

First general elections in the country were held only a year ago and the promise of democracy, development and equality was a great source of inspiration for students moulding their aspirations for whom sky appeared to be the only limit. One could also see students professing divergent ideologies and giving expression to their views passionately.

For all these reasons—quality of minds, cosmopolitanism, rising aspirations and freedom of thought and expression—the interaction among students in the corridors of the college appeared no less rewarding than their learning in the classrooms. Some of the conversations could prove to be quite far-reaching influencing many a career.

It was my privilege to have been a contemporary of a galaxy of students in the College who made a mark later in their chosen professions. Shyam Benegal, who did his Masters in Economics at the College, chose to get into making films—not-so-dry a field like economics. Because of his originality, imparting social relevance and depth, his career in film-making earned him national recognition. For many of us in the College, this quality in him was totally unsuspected. It is clear, in retrospect, that he was highly sensitive to the social milieu around Hyderabad in the Telangana villages in the not-so-distant feudal era.

R. Vaitheswaran—President of the College Students' Union for 1954-55 (when I was the Vice-President)—with a leftist background, having been jailed for some time, was a brilliant scholar-orator, who could hold his audience in rapture. He moved to the USA and has been teaching Economics there. There was K. Rajagopalachari, a silver tongued orator and scholar, who brought laurels for the College and enriched the intellectual life. Basheeruddin Ahmed—President of the College Students' Union in the following year—who did his M.A. in Political Science from the College, joined the Delhi's Centre for the Study of Developing Societies and pioneered empirical studies on the democratic participation of the people through surveys on their voting behaviour and, thus, in Ashis Nandy's words "held a mirror to Indian democracy". He rose to become the Vice-Chancellor of Jamia Millia Islamia in Delhi. Waheeduddin Khan, his younger brother and my class-fellow, who specialised in Regional Planning and Agriculture, became the Founder-Member and first Director of the Centre for Economic and Social studies (CESS), Hyderabad.

Among our teachers, Prof P.V. Rajgopal was most popular. He was sharp and could command rapt attention from students because of his lucidity of exposition and wit. One could never think of missing his class in Political Science. Thanks to the communication skills of Prof. (Madam) M.F. Jussawalla, Principles of Economics became interesting and absorbing. A good combination of scholarship and communication skills in a teacher made the classroom interesting and rewarding, especially for students at the graduate level.

I gained a great deal from the teachers as well as fellow students. In retrospect, I feel I could have perhaps gained more if there was an appropriate balance between my pursuit of studies and my commitments as the General Secretary, All Hyderabad Students' Union, affiliated to the All India Students Federation (AISF). In fact, my joining the College after 4 years of interruption was prompted as much by the desire to pursue studies as by the urge to make use of this opportunity to build up the student movement. Clearly, because of my virtual whole-time involvement in the student movement, there was a conflict of objectives. This is how, despite my opting for Economics as the main subject, in practice it became subsidiary, and practical politics continued to be my main vocation. This would not have mattered much if later in life one chose politics or business as a profession. But having chosen academic career—doing Ph.D. at the Delhi School of Economics, followed by a research assignment at the Institute of Economic Growth (IEG), Delhi—one could see the gaps in one's equipment, which one could make up only by running fast by withdrawing oneself from several other desirable pursuits. Yet, when it comes to studies, one can never fully make up for the time lost at a young and impressionable age.

Nizam College News Letter, May 2012 (Special Issue: Celebrating 125 years).

43

All India Students' Federation Conference, 1996

NEARLY 43 years ago, that is, in the winter of 1952, we, under the auspices of the All Hyderabad Students' Union, had the privilege of hosting the Annual Conference of the All India Students' Federation (AISF) at the Nizam College grounds in Hyderabad. It was a matter of great pride for me and my colleagues as humble workers of the AISF to host this Conference. I understand from my friends that it is after the lapse of 43 years that the National Conference of the AISF is being held in Hyderabad. We, therefore, feel greatly honoured to be the hosts again for this rare event.

Hyderabad city has been known for its hospitality. Still cosmopolitan, the life in the city reflects secular traditions with people speaking various languages and professing different faiths. Even in the most volatile days of the 1940s and the 1950s, the city was singularly free from any animosities and tensions between different religious and linguistic groups. As you move in the city, I am sure, you will be able to communicate with different sections of people freely and with ease.

Andhra Pradesh has a rich heritage of progressive student movement. Before the formation of Andhra Pradesh in the mid-1950s, there were two streams of this movement, one from the Telangana which formed part of the old Hyderabad state and another from the rest of Andhra Pradesh. Being the capital of old Hyderabad state and the nerve centre of the Telangana region, the city had witnessed some of the glorious struggles led by the All Hyderabad Students' Union which was affiliated to the AISF. During the 1940s, the students in Telangana were in the forefront of the movement for civil liberties and for the liquidation of the princely state. Thousands of students suffered victimisation, torture, imprisonment, and several of them laid down their lives.

After the achievement of independence, the student movement in the country became increasingly concerned with issues like universalisation of education and the expansion of employment opportunities. It is gratifying to note, however, that the student movement led by the AISF and other like-minded organisations did not take a narrow view of these problems which, immediately, affect the students. The movement has been aware that these goals can be achieved only within a framework of sustained and equitable development. Further, the student movement led by these progressive organisations has remained steadfast in its commitment to the goals of secularism and socialism. Indeed, education has rightly been regarded by them as a means for achieving social change on secular and egalitarian lines.

The younger generation is best fitted to carry this noble message to the rest of the society. It is also best equipped to build the society on these healthy lines. Therefore, there is a sense of revulsion among the students against any moves to undermine the role of the state in making education accessible to those lacking in necessary means. Similarly, the student community is understandably unhappy over the pattern of development that is not sufficiently broad-based and is unable to secure social justice. Their struggles for increasing the number of freeships and scholarships for students belonging to the low-income groups and the disadvantaged sections of the society are fully justified. Rather, we should feel sorry that the students are forced to undertake such struggles for meeting the basic social needs even today when we are completing half a century of our independence.

While the student movement is in the forefront of the struggles for a new social order that is just and equitable, it has simultaneously been busy building up human resources in the widest sense of the term: it has been keen to build up the new generation which is technically proficient, socially committed and adequately trained for enhancing the quality of the functioning of our democratic polity. Among the activists of the AISF, we have seen quite a few who achieved academic excellence, have been imbued with social commitment and made a mark in whatever profession they chose.

In a highly complex and fast changing world of today, such qualities become even more challenging for the younger generation. Recent experience has shown that development patterns and paradigms chosen earlier have had to undergo drastic changes in response to the changes in technology and in the systems of management. Yet, deviations from, or compromises on, certain basic values have often landed people and the whole nations in trouble. Firm commitment to the basic ideals needs to be combined with great deal of flexibility in the choice of the exact framework or means for achieving the goals. This calls for attitudes like openness of mind, respect for differing points of view and readiness to face criticism without attributing motives.

Welcome Address, as Chairman, Reception Committee, at the 24th National Conference of the AISF, February 7-9, 1996, Hyderabad.

44

Professor V.K.R.V. Rao

Institution Builder Par Excellence

VIJAYENDRA KASTURI RANGA VARADARAJA RAO was truly a colossus in the world of social sciences—a brilliant economist with outstanding contributions towards the estimation of national income of India and analysis of the relevance of Keynesian economics to the understanding of the problems of developing economies and a policy-oriented economist who actively participated in several high-level expert bodies.

He was an institution builder par excellence. A person of boundless energy, he founded the Delhi School of Economics and, subsequently, the Institute of Economic Growth (IEG), at the University of Delhi, and the Institute for Social and Economic Change in Bangalore. He was instrumental in establishing the Indian Council of Social Science Research and a number of Agro-Economic Research Centres in different parts of the country. These institutions have been providing opportunities for innumerable scholars from diverse disciplines in social sciences to grow and contribute to the understanding of the development process in India.

While V.K.R.V. Rao's contributions as a pioneer in Indian economics and as an institution builder are widely recognised and documented, very little has been written about his intellectual personality—his concerns, passions, motivations and attitudes. He was deeply patriotic and fired with the ambition of carving out a role for professional economists and social scientists in general, in the great task of nation-building in independent India. He was a Nehruvian in his outlook and remained so till the end of his life. Even in the busy days of institution-building, he was alive and highly sensitive to major sociopolitical events in the country, on which he wrote and spoke extensively and inspired his colleagues to react and contribute. His passion for institution-building cannot be explained except in terms of his deep and abiding interest in the problems of Indian economy, society and polity. His intense desire to keep in touch with the latest developments in these spheres was transparent till the end of his life.

Attracting the best available talent to his institutions was his craze which led to his occasional disregard for rigid rules and regulations in the matter of appointments and promotions. It used to be said that Dr Rao can be accused of anything but not of being parochial. He took particular care to ensure that in his institutions, scholars from various regions and with different social backgrounds were appointed. He was even keen to accommodate scholars professing different ideologies and could enjoy with a sense of detachment the battles between contending ideologies. He would as readily recruit scholars trained in Indian universities as those from first-rate centres of learning abroad. Balancing between scholars belonging to different disciplines in social sciences, and between the young and the old, he sought to provide breadth of approach while injecting dynamism in his institutions.

He constantly exhorted his colleagues to devote their time not merely to individual research pursuits but also to group endeavour in research on problems of national importance; to spend adequate time on training, particularly on building up the young scholars; and to readily communicate the findings of research to students, teachers, policy-makers and the public at large.

His overbearing personality and his impatience at deviations from the norms set by him may have displeased some, but not for ever. He was too simple, straight and exacting—more exacting on himself than on others. Though temperamental, he was a democrat at heart, for he was refreshingly sensitive and responsive to the friendly advice given to him. His uprightness was unmistakable.

He appeared to look at economic reforms and changes in economic policy with an open mind. But he continued to be firm in his commitment for social justice and hence his emphasis on employment generation, rural development, and human and social development in general. This was amply revealed in his writings and speeches even during his closing days.

IEG at Fifty—Recollections, Retrospect and Prospect, Institute of Economic Growth, Academic Foundation, New Delhi, 2008.

45

Professor Sukhamoy Chakravarty

An Erudite Scholar and Outstanding Planner

IT was a great privilege for me to have been associated with Professor Sukhamoy Chakravarty for over two decades. I first knew him as a scholar and then came into contact with him when he was active as a planner in the Yojana Bhawan. All this, I would say, was the first phase. I would regard the 1980s as the second phase in Sukhamoy Chakravarty's career as a planner, a policy-maker and a promoter of social science research in India. During this last decade, I had the privilege of working closely with him as a member of the Economic Advisory Council to the Prime Minister, the Research Advisory Committee of the Planning Commission and the Indian Council of Social Science Research (ICSSR)—all under his chairmanship.

As a planner in the 1970s he was relentlessly translating the goals of self-reliance and social justice into plan models and programmes. In this role he appeared as a crusader. I think the 1980s was equally a period of turmoil for Sukhamoy Chakravarty. This is because he was not a dogmatist who could comfortably stick on to the old ideas and spare himself the struggle of exploring the new and the relevant. Nor was he given to the easy ways of jumping from one position to the other. His was an intellectual struggle of a serious, constructive and an honest scholar. This perhaps provides the clue to the intense strain that his already fragile physique was subjected to in the 1980s.

He was extremely cautious and balanced in his evaluations which were characterised by the sweep of his understanding of the Indian realities in their varied dimensions as well as a sense of commitment to the goals of self-reliance and social justice. A scholar of wide-ranging interests as he had been, he displayed a rare breadth of outlook and humane approach as a policy advisor and leader of the academic community. Among the scholars, policy-makers and administrators with whom he came into contact, he could see the strength of each one of them and carry them with him regardless of their ideological positions.

In the passing away of Sukhamoy Chakravarty the country has lost not only an erudite scholar but an outstanding planner and policy-maker, and a fine human being.

Mainstream, September 15, 1990.

46

Professor P.C. Mahalanobis and Sixth National Statistics Day

IT is indeed a great pleasure and honour to be present on the occasion of the birthday of Prof Prasanta Chandra Mahalanobis, being celebrated as the National Statistics Day. Prof Mahalanobis was the founder of the Indian Statistical Institute and helped the government in building up a well-designed national statistical system by instituting the National Sample Surveys (NSS), Central Statistical Organisation and State Statistical Bureaus.

With encouragement from Pandit Jawaharlal Nehru, the then Prime Minister of India, he drafted the Second Five Year Plan in the mid-1950s. The mathematical framework underlying this Plan became famous as the Mahalanobis Model—also as the Nehru-Mahalanobis Model. This exercise inspired several young economists to take interest in economic planning and development of the country and make it a mission in their lives.

For example, Prof Sukhamoy Chakravarty, in his memoirs included in the collection of his papers which Amit Bhaduri, Raghavan and myself had edited, says that as a graduate student in the mid-1950s in Calcutta, he was inspired by the Nehru-Mahalanobis Model and pursued the study of economics with an ambition to contribute to economic planning and development of the country. It so happened that within less than two decades, i.e., in the early 1970s he was appointed Member of the Planning Commission. He gave his best to the Planning Commission by putting in a great deal of effort and time for plan modelling using the vast amount of statistical data available.

The Mahalanobis model was well received by development economists in general because of its accent on achieving self-reliance in industry and technology by giving high priority to heavy industry and development of scientific laboratories. There was hardly any infrastructure worth the name in the early stages of development of the country in the 1950s. The strategy of self-reliance practised under the Nehru-Mahalanobis model enabled building up infrastructure on a significant scale. The acceleration in the gross domestic product (GDP) growth following liberalisation of economy in the early 1990s owes very much to the development of physical and social infrastructure in the pre-liberalisation period under the strategy of self-reliance.

Organised statistics are necessary for making proper decisions in face of uncertainty. The ruling elites the world over made their momentous decisions for centuries, without the help of organised statistics, simply on the basis of common sense, impressions, guesses and even religious beliefs and prejudices. But the modern world is highly complex and interrelated, so the risks of ill-informed individual decisions are great for the system as a whole. In the modern age, by using well-collected statistical data, governments make their short-term and long-term plans and projections for economic development and human welfare. Similarly, business enterprises and corporations are increasingly using such data for their projections, evaluations and mid-course corrections.

My exposure to statistics started with my work for PhD at the Delhi School of Economics in the late 1950s. Earlier, I had imbibed several ideas on the economy and development through text books and even radical ideology, as the young are usually prone to. Though held strongly, my ideas, opinions and prejudices, for example, on land distribution, agricultural productivity and household consumption were yet to be subjected to scrutiny by the hard data available. In a way, a beginning for such testing was provided by an extensive field survey of farmers I did for the purposes of my PhD, in East Godavari, Krishna, Ananthapur and Karimnagar—my home district. This was done through my personal participation by staying in the villages and through reliable local assistance with a view to ensuring the quality of data collected. This very process cultivated in me respect for data and readiness to test the ideas on the basis of data.

It may be of some interest for the young friends present here to know my own experience with the modification or even rejection of some of my strongly held notions when I came face to face with the relevant data. Before confronting with the relevant data, my ideas on land reforms were rather simplistic in the sense that redistribution of land from the rich landed sections to the landless poor would, by itself, be sufficient to eradicate rural poverty. But after a full exposure to the data on the amount of surplus land to be made available after the imposition of ceiling on land holdings and the number of landless poor as well as the marginal holdings needing land, it became clear that it would not be possible to lift the rural poor above the poverty line merely by redistributing land. Thus whereas land reforms were necessary, it would not be possible to eradicate rural poverty unless productivity of land is increased substantially through technological change. Since then my research work came to be focused on technological change in Indian agriculture.

Drawing inspiration from the collective and co-operative farms in Russia and China, many economists in India in the 1950s and 1960s thought that co-operative farming is essential for raising agricultural productivity by reaping economies of scale. Even before analysing the data, my exposure to the field surveys in the villages made me skeptical about the feasibility of such a proposition in the Indian rural social context and even about the possible advantages of large-scale farming. But very soon the analysis of my survey data as well as the Farm Management Studies conducted by the Directorate of Economics and Statistics, Government of India, in several parts of the country clearly established that small family farms were more efficient, as output per acre among them was much higher than that among large farms. This finding was so robust that the attention soon shifted to theoretical work explaining why small farms were more productive than the large ones. Prof Amartya Sen initiated the debate on this in the mid-1960s through an article in the *Economic and Political Weekly*. Several others, including myself, joined this debate. In the light of this emerging statistical evidence, the talk on the need for co-operative and collective farming in India simply evaporated much before the collectives and co-operatives were disbanded in Russia and China.

The third example relates to my notion on the consumption of cereals. I strongly believed that the energy requirements of the poor, because of hard manual labour they have to put in, are higher and so they consume more cereals per head than the rich. When Dr R. Thamarajakshi, a contemporary of mine doing PhD on Consumption and Savings by using NSS data—who was later to become secretary of the Department of Statistics, Government of India—told me that per capita consumption of cereals is in fact higher among the richer households than among the poor, I thought there was something wrong with the data being used. After a scrutiny of NSS data and discussions with certain senior and experienced scholars, I had to fall in line and accept the finding. The basis for my belief was the error of equating the need for higher consumption of cereals by a farm labourer with his ability to consume such quantity.

For most of my working period. I have been basically a user or 'consumer' of data. My association with the processes of data generation and organisation started virtually at the time of retirement from my job in 1992 when I was invited to join as Chairman of the Advisory Committee on National Accounts Statistics. My predecessor Prof V.M. Dadekar was an eminent statistician. My only credentials were of a user of data in the field of agriculture and development planning in general. But this new acquaintance with problems of data generation, its collation and presentation after making necessary adjustments, etc., made me increasingly conscious of the limitations of data presented. I almost felt that I would not have had the confidence to draw many of the conclusions that I did in my research career if I was aware of some such data limitations. It is almost like enjoying a meal in a five star hotel but abruptly giving it up after getting a chance to peep into the kitchen and having a glimpse of how clean the environment there is! However, one should be aware of data limitations which should lead one to be cautious in their interpretation rather than abandoning the baby with bath water. Planning with imperfect data is any day far superior to planning with no data at all.

Remarks made on the occasion of the 6th National Statistics Day Celebrations (on the birthday of Prof P.C. Mahalanobis), and Fourth Statistics Olympiad; Dr C.R. Rao, Advanced Institute of Mathematics, Statistics and Computer Science, University of Hyderabad, 29 June, 2012.

47

From PhD Student to Board Chairman

Experience at IEG

IN January 1958, I was selected for a PhD Fellowship in the Economic Development Section at the Delhi School of Economics (DSE), supported by the Ford Foundation, for undergoing training in Research Methodology for a year before I was registered for PhD. For my thesis, in addition to Prof. Vikas Mishra, under whose supervision my work was done, I was lucky to get guidance from Prof V.K.R.V. Rao and Prof A.M. Khusro.

While I and my contemporaries started working for PhD with enthusiasm, many of our seniors already working for PhD at the DSE had spent several years—some nearly a decade—causing some pessimism among us. Apparently, their interaction with supervisors was inadequate. The situation changed soon with the successful completion of a series of PhD theses in the 1960s, notably under the supervision of Prof Khusro. Prof V.K.R.V. Rao made it a point since then to mention that Prof Khusro gives enough time to his students.

By the time I started working for my PhD, I had made up my mind to specialise in Agricultural Economics, as agriculture on all accounts appeared critical at that time for overall growth of the economy and eradication of poverty. My rural background and acquaintance with the agrarian movement prompted me to seek answers to the many questions facing the peasantry through research and learning. Moreover, for 12 years since the mid-1940s, I was active in the student movement led by the Left, aimed at sociopolitical transformation, beginning with the campaign for civil liberties against the Nizam's government, which led to my rustication and arrests, causing frequent interruptions in my studies. Thus while my past did provide an inspiration and a sense of relevance and meaning in my research, it also brought me face to face with the need to improve my equipment in economics and also 'unlearn' quite a bit on the ideological plane if research is to be pursued with objectivity free from pre-conceived notions and other prejudices.

Thus my PhD work became a practice in 'running fast' to make up for the lost time. It cultivated in me respect for facts and readiness to question the settled ideas if facts so demanded. It also turned out to be an exercise towards precision in writing.

When the Agricultural Economics Research Section was started in 1961 at the Institute of Economic Growth (IEG), I was appointed as a Junior Fellow. After a few years, the authorities at the IEG thought of encouraging me to spend a year or so at some good university abroad. When the suggestion came to me in early 1966, instead of seeing in this gesture an opportunity for learning, I tended to detect a 'craze' for 'foreign credentials'! But it did not take much time to see the positive side, although shedding prejudices totally was to wait until I spent a few months in the library at the Department of Economics, University of Chicago, and completed auditing courses offered by stalwarts like T.W. Schultz, Milton Friedman, George Stigler and others. My wife Krishna was with me, herself working in the South Asia Library and providing me, as always, full support for my academic pursuits.

The number of papers and books one could read in a year sitting in a corner in the library was simply unimaginable even in a decade in India. I could also complete my paper on 'Sharecropping' which later appeared in the *Journal of Political Economy*. I felt so much stimulated, even excited, that I thought of staying on for a few months more. Accordingly, I wrote to Prof P.N. Dhar, then Director, requesting him to extend my fellowship for another three months. This he did graciously, but with a rider that I should not ask for further extension!

A decade later, the IEG negotiated with the Ford Foundation to revive the programme for supporting its scholars to pursue post-Doctoral work abroad. A number of scholars availed of this and several of them, including Professors B.B. Bhattacharya, N.S. Siddharthan and S.C. Gulati, continued to work and make outstanding contributions at the IEG. The IEG's approach to scholars taking up academic assignments abroad for limited periods has been positive so long as they did not adversely affect the Institute's own research and training activities.

Declining to accept long-term or regular assignments with international institutions is in line with Prof V.K.R.V. Rao's own practice. Such assignments, apart from depriving the country of much needed talent, carry the risk of alienating the scholars from the developments in the country leading to their analyses and prescriptions becoming sometimes out of tune with the situation in the country. On the other hand, exposure to the wider intellectual currents in the world at large through stay and interactions for limited periods abroad can be refreshing and rewarding for fruitfully addressing the country's problems by liberating oneself from the 'frog-in-the-well' mind set.

If most of the IEG Faculty preferred to remain at the IEG by declining offers from within the country and abroad, it is not only due to their commitment to the above values but also to the 'work culture' at the IEG. Freedom of thought and expression, a sense of social relevance in research and academic excellence, are the foremost values nurtured by Prof V.K.RV. Rao. These have sustained us intellectually and spiritually over the decades. The IEG's location in Delhi, adjacent to the DSE in the University campus, conferred unique advantages. Many from the Faculty had completed their PhD from the DSE before joining the IEG. These and others have continued their beneficial interaction with the DSE, even as the life at the IEG remained quiet when the University was in turmoil. Easy interaction with visiting scholars from abroad, effective exposure to major policy initiatives and participation in the expert committees at the national level are enriched further by the insights from the field work by a number of IEG scholars, notably from the Population Research Centre. Accountability for their research work was sought to be ensured basically through the presentation of their work, in several stages of the project, at faculty seminars where experts from other institutions were also invited.

A National Seminar on Relevance in Social Science Research was organised at the IEG in 1980 to clarify and sharpen our understanding of the relevant issues. This resulted in an important publication from the IEG. Again, in the mid-1990s, there was an interesting debate at the IEG on the need for its research agenda to respond to the changing economic environment. Through the flood of emerging literature, exchange of visits and academic interactions, the IEG scholars are sufficiently exposed to the on-going intellectual currents in the world and are amenable to changes in thinking to the extent warranted. It is gratifying that the research community at the IEG not only responded positively to the changing economic environment but also evolved, on its own, a balanced approach to economic reform much before the view 'economic reforms with human face' or 'inclusive growth' came to be officially accepted.

An issue needing exploration is the prospect for multi-disciplinary research. At the moment, when institutions like the IEG claim to be multi-disciplinary, it simply means scholars from different disciplines sitting and working under the same roof! They do exchange views occasionally, but on matters mostly outside their disciplines. Joint studies undertaken, for example, by sociologists, economists and demographers on subjects of common interest could perhaps serve as good experiments for seeing their usefulness and for advancing the collaborative efforts further.

In the early 1970s, Prof V.K.R.V. Rao took the initiative, by involving some of us, to restructure the management of the Institute with a view to making it democratic and self-governing through the participation of the senior faculty at various levels. Accordingly, the rules and regulations of the Institute were amended in several rounds to ensure, among other things, greater representation to the faculty in the Governing Body and the Committee of Administration and Finance. The Academic Committee takes decisions on all academic matters. A number of other committees have also been functioning well. Looking at the functioning of the Institute as Chairman, I find that the IEG has evolved, over a period of time, procedures and practices that are well-settled by now with the widest possible participation by the faculty, without any populist overtones. The IEG has indeed matured.

An indirect but important contribution of the Institute, which can not find a place in the usual list of its achievements, is the education of the children of the faculty and other staff. These children—in many cases extremely well-endowed—are spread over in different parts of the country and the world, making a mark in their fields. The IEG can legitimately take pride in such human capital. No less significant is the social capital, for example, the successful functioning of the IEG Employees' Cooperative Credit and Thrift Society, Staff Welfare Committee and the good-will among different sections of the IEG community.

Published in *IEG at Fifty—Recollections, Retrospect and Prospect*, Institute of Economic Growth, Delhi, Academic Foundation, New Delhi, 2008.

48

Padma Bhushan Award Felicitation

My Response

I am so much touched and overwhelmed by the affection shown to me and the feelings expressed by the speakers this evening that I do not find adequate words to express my gratitude. It is customary on such occasions to enumerate the achievements and qualities of the person being felicitated. In my view, this is only one side of the coin. If the person concerned is honest and introspective, he would easily discover that many of the 'achievements' attributed to him are the product of accidents or confluence of circumstances or sheer luck. I do not believe that these events are pre-ordained and so I do not read any supernatural meaning in them. But the fact remains that they are the product of circumstances very much extraneous to and beyond the control of the individual concerned. This is what I see as the other side of the coin.

To narrate an instance, when I joined the Nizam College in 1953 for the graduate course in humanities after an interruption of 4 years, I had to choose the main and subsidiary subjects. The 'main' subject chosen would determine future specialisation at the M.A. and Ph.D. levels. Impressed by the smart city-bred fellow students, much younger than me, speaking English fluently with British accent, because of their schooling through English medium, I decided to opt for English Literature as my main subject and Economics and Political Science as subsidiary subjects. Then, just one or two days before the last date for the choice of subjects was to elapse, I accidentally met in the corridors of the Nizam College an intimate friend and colleague of mine in the student movement, Shri Burgula Narsing Rao, to whom I mentioned about the choice of my subjects. He immediately reacted and told me that Economics has a better future and it is also not correct to think that by taking up English Literature one can really improve one's English! Somehow, I was persuaded by his reasoning and the next day, i.e. a day before the last date was to be over, I changed my main subject from English Literature to Economics.

A few months later there was an elocution competition in Urdu, at the Urdu Hall, Himayatnagar, on 'Unemployment Problem'. I was interested in such issues and as a student of Economics, I started picking up the relevant statistics. It was an exciting moment for me to successfully participate in this competition. One of the judges happened to be Dr A.M. Khusro whom I met there for the first time. He had just joined the College of Arts of the Osmania University as a Lecturer after completing his Doctorate from the University of Leads, UK. I was inspired by his summing-up speech and encouraged by his kind words about my performance in the debate. Since then I kept in touch with him. After doing B.A. from the Nizam College, I and my class fellow Waheeduddin Khan, who was later to become the first Director of the Centre for Economic and Social Studies (CESS), shifted to the College of Arts for doing M.A. The next year Dr Khusro was appointed as Professor in Economics at the Delhi School of Economics (DSE) by Prof V.K.R.V. Rao. Prof Khusro encouraged me to do my PhD from the DSE. Luckily, just about that time, I had an opportunity to meet and interact with Prof V.K.R.V. Rao at a World University Service Seminar held at the College of Arts, Osmania University. He also encouraged me to do PhD at the DSE. These fortuitous contacts and encounters with such eminent teachers opened up opportunities for me in the profession.

I learnt precision in writing from Prof. Vikas Mishra who was my supervisor for PhD. It was fashionable in those days to submit bulky theses. But, Prof Mishra used to say one should always ask whether the sentence that one has written is really 'necessary'. After reading the first draft of my thesis, he suggested deletion of a number of passages by placing brackets indicating the portions to be deleted. On one occasion, I felt that there were too many brackets placed by him, and after deleting all that material, very little would remain. So, I went to him and started arguing as to why many of these sentences needed to be retained. He very patiently heard me and quietly got up, picked up from his shelf an old draft of his own Ph.D. thesis and showed me the similar brackets placed by his supervisor, Prof. Arthur Lewis of the Manchester School of Economics, the author of the famous paper 'Economic Development with Unlimited Supplies of Labour', who was later to be awarded the Nobel Prize in Economics. This cooled me down and persuaded me to get rid of some of the 'unnecessary' material. In the course of the preparation of our own book on Andhra Pradesh Development, my colleagues at the CESS would have frequently encountered such brackets placed by me on their drafts. I am sure many of you will pass on this tradition to your own colleagues and students.

At the Institute of Economic Growth (IEG), Delhi, I had the good fortune of interacting with a number of learned colleagues over a long period of time. For example, even before sending my very first book at the IEG on Production Functions to the press, I had an opportunity to take help from Prof K. Krishnamurthy for giving final touches by incorporating certain statistical tests with which I was not quite familiar then. Prof. Krishnamurthy already mentioned about the intellectual influence that Prof. Dharm Narain had on me. Among other things, he used to say, "never rush to the press when you think your paper is complete. Better keep it in a drawer for a few weeks and look at it again for carrying out possible revisions".

I was on leave from the IEG for quite some time intermittently to work on several commissions and committees. One's own contribution to the deliberations of these committees apart, one learns a great deal while working on such expert committees. No one individual can destroy big committees and institutions, but one can easily destroy one's own standing and reputation by one's manner of functioning. I was lucky to have scraped through these assignments, thanks, not the least, to the understanding and co-operation from colleagues on such committees.

During the nearly 12 years since I returned to Hyderabad after my retirement from the IEG, I have been pursuing my work at the CESS on the post-reform scenario on agriculture, poverty and environment. My working at the CESS, including as Chairman, has been totally free from any stress. I have enjoyed the work as a co-editor of the volume on Andhra Pradesh Development brought out recently. Thanks to the full co-operation extended to me at the CESS by the authors of papers, this experience has been both educative and ennobling.

I would like to finally express my deep sense of gratitude to all of you present here for the regard and affection shown to me through this felicitation.

In response to the felicitation at the Centre for Economic and Social Studies (CESS), Hyderabad, on 6 February, 2004, on being awarded 'Padma Bhushan' by the President of India.

49

Continuity and Change

Response to the Felicitation at the C.R. Foundation

WHEN Com. Y.V. Krishna Rao garu met me with a proposal to hold this felicitation function under the auspices of the Neelam Rajashekhara Research Centre, I was very much touched by this gesture and naturally my thoughts went down the memory lane when I was actively involved in the student movement for 12 years from 1945 to 1957, virtually as a whole-time worker. In 1957, my career as a full-time researcher in Economics started with my admission to Ph.D. in the Delhi School of Economics (DSE). The question that comes to my mind is whether this amounted to a 'clean break' from my past. Obviously, there was a break in my career, as I was now engaged in a different field of activity altogether. Yet, on reflection, I realise that I did not really start with a 'clean slate' and there have indeed been strong elements of continuity in my work in terms of overall goals, commitments, approaches and attitudes.

The very first element of continuity I find is the urge for freedom of thought and expression for which we struggled hard during the freedom movement. My refusal to resign from the All Hyderabad Students' Union and to tender apology to the Nizam Government for participating in 'political activities' represented an assertion of this right to freedom of association, thought and expression. Fortunately for me, I found a place at an autonomous institution like the Institute of Economic Growth (lEG) in Delhi, which was founded by Prof V.K.R.V. Rao, who himself was inspired by the freedom struggle and who believed in scholars with diverse ideological commitments pursuing their research work with full freedom of thought and expression. Likewise, I am now lucky to be associated with the Centre for Economic and Social Studies (CESS), Hyderabad, which was founded by Shri B.P.R. Vithal with the same objective of promoting research work in social sciences with full freedom of thought and expression. Today, at the IEG or at the CESS, or for that matter at any reputed social science research institution in the country, no one can say that the ideas expressed in one's writings are not truly one's own or that one would have written differently if there was no 'pressure' from the 'authorities'. I am glad that Com. K.L. Mahendra, whom I have known closely from my student days, has rightly underlined the importance of objectivity in research and presentation of one's findings in a free and fearless manner. The tendency to do motivated research only to provide support to received theories, as he rightly says, will not be beneficial in the long-run to the very same people whom the researcher is keen to help.

The second trait I imbibed from my participation in the student movement of those days is the secular outlook. I remember in the mid-1940s when I was the secretary of the Karimnagar Unit of the All Hyderabad Students' Union, Hindu and Muslim students used to be roughly in equal number as our members. We used to keep aloof from, and in fact fight against, the caste-based associations of students. One may ask: What has secular outlook to do with research work in Economics? I think, economist being a citizen, first and foremost, secular outlook makes for a healthy citizenship. But, I strongly believe, after knowing several scholars from close quarters, that a social scientist with a narrow outlook on life is incapable of producing outstanding work of broader dimensions in respect of its social implications.

The third trait for which I value my formative 12 years in the student movement is the spirit of nationalism. My teacher and mentor Professor V.K.R.V. Rao had set the example for the younger scholars when he declined to take up lucrative assignments abroad and chose to work within the country. While I was following his foot steps, my source of inspiration goes back to the student movement of the 1940s and the 1950s. This inspiration from the past prompted me to travel extensively in the country, acquaint myself with the problems at the grass roots and do research on them.

The fourth quality, which perhaps had a more pervasive impact on my research work than any other, relates to the issue of relevance in setting up my own research agenda. As I said earlier, we do enjoy freedom in pursuing research work. But what research agenda you set for yourself depends very much on your academic interests, social commitments and inclinations. For 12 long years up to 1957, the shared goal was to bring about social transformation through the political movement. For various reasons, 1957 was a watershed in my career. It suddenly dawned on me that, given my personal circumstances, I can pursue the same goals more effectively through research in Economics and Economic policy. So, in a way, the social agenda was already given and handed down to me from my past. This continuity in my overall commitment and inspiration has sustained me to this day intellectually and spiritually. This has taken me to explore areas like agricultural economics, rural development, poverty alleviation, problems of small and marginal farmers, agricultural labour, their employment and wages, regional disparities in development, etc.

There is the fifth attribute which continued in me from my past but about which I was not conscious and was virtually blind until several of my well-wishers drew my attention towards the late 1970s when I was the Director of the IEG. For nearly 20 years after I left the student movement, I was confined to my room and to my studies and very few people thought that I would be able to take up the new challenge as the Director of the lEG which involved lot of administrative work and dealing with people of diverse temperaments. Immediately after the Emergency was lifted, the University of Delhi was in turmoil and when after some initial troubles, lEG saw peace returning to its campus even as the rest of the University was still in trouble, Dr V.K.R.V. Rao remarked to me, "Your Student Federation days are helping you"! After that I have had the opportunity of working in several expert committees and commissions either as a member or as chairman. In one such Commission that I chaired (National Commission on Rural Labour), there were at least half a dozen members of parliament belonging to parties with diverse political ideologies, e.g. Congress, Janata Dal, CPI and CPI(M). I was singularly lucky to scrape through such committees and commissions, thanks, not the least, to the understanding and co-operation extended to me by the members. But I am also indebted to my pre-1957 years known for the single-minded pursuit of social goals by interacting with persons of diverse ideologies and temperaments.

When I am talking about the past, I must mention about the help I received from Com. Chandra Rajeshwara Rao garu and Com. Neelam Rajashekhara Reddy garu, after whom this research Institute has been named, towards carrying out field work for my Ph.D. thesis, in West Godavari, Krishna and Anantapur districts. Through the contacts given by them, I was able to collect data which was highly reliable and therefore added to the quality of my research work. I remember my discussions with Shri Neelam Rajashekhara Reddy garu way back in 1959 in Anantapur when I was doing my field work. He was frank about the gaps in our understanding of agrarian situation in the state and hoped that my research would help to bridge some of these gaps. I must also mention about my field work in Digavelli village in Nuzvid *taluka*, where I had the privilege of staying with the parents of Com. N.R. Dasari garu. A quarter century later I happened to visit Krishna District when I was a member of the Planning Commission, and I could not then resist the temptation of visiting that village and the home where I stayed.

I would like to express my deep sense of gratitude to the speakers this morning and to you all present here, for the regard and affection shown to me through this felicitation.

Response to the felicitation at the N.R.R. Research Centre, C.R. Foundation for Social Progress, Kondapur, Hyderabad, on 14 March, 2004, on being awarded 'Padma Bhushan' by the President of India.